MW01119476

ᐱᓱᖅᑲᑎᒋᓪᓗᒍ ᐋᓚᓯ
Walking with Aalasi

Inhabit Media Incorporated, Toronto and Iqaluit
A Nunavut Bilingual Education Society Project

Printed in Canada.
ISBN 978-0-9782186-7-6
Joamie, Aalasi and Anna Ziegler

Walking with Aalasi: An Introduction to Edible and Medicinal Arctic Plants
Aalasi Joamie and Anna Ziegler.
Published by Inhabit Media in collaboration with the Nunavut Bilingual Education Society with funding from the Department of Culture, Language, Elders, and Youth and Canadian Heritage. Text in English and Inuktitut.

Editor	Neil Christopher
Authors	Aalasi Joamie and Anna Ziegler
Translator	Rebecca Hainnu
Researchers	Aalasi Joamie, Rebecca Hainnu, and Anna Ziegler
Illustrator	Patrick Little and Danny Christopher
Photographer	Anna Ziegler
Cover Designer	Ellen Ziegler
Copy Editors	Louise Flaherty (Inuktitut) and Shelley Ross (English)

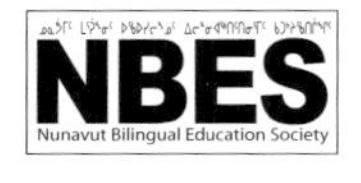

Canadian Heritage   Patrimoine canadien

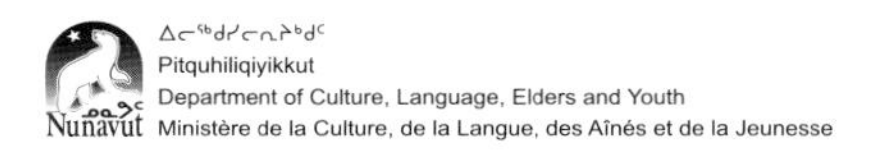

ᐱᓱᖅᑲᑎᒋᑦᓗᒍ ᐋᓚᓯ

ᑕᑯᑎᑦᑎᒋᐊᖖᒐᕐᓂᖅ ᐱᕈᖅᑐᓂᒃ ᓂᕆᔭᒃᓴᓂᒃ ᐃᓅᓕᓴᐅᑎᐅᓯᓐᓇᖅᑐᓂᒡᓗ

Walking with Aalasi

An Introduction to Edible and Medicinal Arctic Plants

ᐋᓇ ᓰᒡᓚᕐ
Anna Ziegler

ᐋᓚᓯ ᔫᒥ
Aalasi Joamie

ᕇᐱᑲ ᕼᐊᐃᓐᓄ
Rebecca Hainnu

ᐸᐅᓐᓂᑦ
Dwarf Fireweed

ᐃᓗᓕᖏᑦ
Table of Contents

ᖁᔭᓐᓇᒦᖅᑕᐅᔪᑦ
Acknowledgements 1

ᐱᓕᕆᐊᑦ ᒥᒃᓵᓄᑦ
About this Project. 3

ᑕᑯᑎᕐᑎᒋᐊᕈᑎ ᐋᓚᓯ ᔫᒥᒦᖖᒎᖅᑐᑦ
Introduction by Aalasi Joamie 6

ᓇᓗᓇᐃᒃᑯᑕᑦ ᓄᓇᕋᐅᑉ ᐊᑑᑎᖕᒋᓐᓄᑦ ᐊᑎᖕᒋᓐᓄᓗ
Summary of Plant Uses and Names 15

ᓄᓇᕋᐃᑦ
Plants 19

ᐳᐊᓗᙳᐊᑦ
Pualunnguat / Arctic Cotton 20

ᒪᓂᖅ
Maniq / Lamp Moss. 26

ᐅᖅᐱ ᓱᐳᑎᓪᓗ
Uqpi Suputillu / Arctic Willow 30

ᖁᐊᕋᐃᑦ
Quarait / Snow-Bed Willow. 36

ᐊᓚᒃᓴᐅᔭᐃᑦ
Alaksaujait / Net-Vein Willow 40

ᐸᐅᓐᓇᑦ
Paunnait / Dwarf Fireweed. 44

ᖁᖑᓖᑦ
Qunguliit / Mountain Sorrel 52

ᓴᐸᖓᕋᓛᖖᒍᐊᑦ ᑐᖅᑕᐃᓪᓗ
Sapangaralaannguat Tuqtaillu / Alpine Bistort. 58

ᐸᐅᕐᖓᐃᑦ ᐊᔾᔨᒌᖏᑦᑐᑦ
Paurngait Ajjigiingittut / Berry Plants 64

ᐸᐅᕐᖓᐃᑦ ᐸᐅᕐᖓᖁᑎᓪᓗ
Paurngait Paurngaqutillu / Crowberry 66

ᑲᓪᓚᐃᑦ ᑲᓪᓚᖁᑎᓪᓗ
Kallait Kallaqutillu / Bearberry. 70

ᑭᒍᑕᖏᕐᓇᐃᑦ ᓇᖁᑎᓪᓗ
Kigutangirnait Naqutillu / Blueberry. 73

ᐸᐅᕐᖓᓂᑦ ᐊᑐᕐᓂᐊᕐᓗᓂ
Using Berry Plants. 76

ᕿᔪᒃᑖᖅᐸᐃᑦ
Qijuktaaqpait / Labrador Tea. 78

ᓯᐅᕋᐅᑉ ᐅᖃᐅᔭᖏᑦ
Siuraup Uqaujangit / Seaside Bluebells 84

ᒪᓕᒃᑳᑦ
Malikkaat / Mountain Avens 90

ᐅᕐᔪ
Urju / Peat Moss 96

ᐃᔨᓯᐅᑎ
Ijisiuti / River Algae 102

ᓂᕐᓇᐃᑦ
Nirnait / Snow Lichen 106

ᐳᔪᐊᓗᒃ
Pujualuk / Dried Mushroom 110

ᐊᑐᒐᒃᓴᒃᑲᓐᓃᑦ
Additional Resources 114

ᑎᑎᕋᕐᑐᕕᓃᑦ
Contributors 116

ᐅᖅᐱ ᓱᐳᑎᓪᓗ
Arctic Willow

ᖁᔭᓐᓇᒦᖅᑕᐅᔪᑦ
Acknowledgements

ᖁᔭᓐᓇᒦᕈᒪᔭᕗᑦ ᐃᑲᔪᖅᑲᑦᑕᓚᐅᖅᑐᑦ ᑲᔪᖕᒋᖅᓴᐃᖅᑲᑦᑕᓚᐅᖅᑐᓪᓗ ᐅᖅᑲᓕᒫᒐᖅ ᐱᔭᕇᖁᓪᓗᒍ.

We would like to acknowledge the contributions and support of the many people who helped us finish this book.

ᐋᓚᓯ ᖁᔭᓐᓇᒦᖅᑐᖅ ᐊᓈᓇᒥᓂ ᓄᓇᕋᐃᑦ ᒥᒃᓵᓂᑦ ᐃᓕᓐᓂᐊᖅᑎᑦᑎᖅᑲᑦᑕᓚᐅᕐᒪᑦ. ᐃᓕᓐᓂᐊᖅᑎᑦᑎᖅᑲᑦᑕᓚᐅᙱᑉᐸᑦ ᐊᔪᕆᖅᓱᐅᔾᔨᓗᓂᓘᓐᓂᑦ ᐋᓚᓯ ᑐᓴᐅᒪᕙᔭᓚᐅᙱᓚᑦ ᐅᑦᑐᒥ ᐃᓕᓐᓂᐊᖅᑎᑦᑎᔾᔭᑎᒋᖅᑲᑦᑕᓕᖅᑕᒥᓂᑦ. ᐋᓚᓯᑦᑕᐅ ᖁᔭᓐᓇᒦᕈᒪᔭᖅ ᐸᓂᑯᓗᒻᒥᓄᑦ ᔫᐋᓂᒧᑦ ᐅᖅᑲᓕᒫᒐᖅ ᓴᓇᔭᐅᕙᓪᓕᐊᑎᓪᓗᒍ ᒥᖅᓱᖅᑎᑦᑎᑎᓪᓗᒍᓘᓐᓃᑦ ᑑᓵᐅᖕᒋᓐᓇᖅᑲᑦᑕᓚᐅᕐᒪᑦ.

Aalasi is grateful to her mother for teaching her about plants. Without her mother's dedicated teaching and encouragement, Aalasi would not have information to pass on to others. Aalasi would also like to thank her daughter Joannie for helping out with interpreting throughout the project and during the sewing lessons when this project was conceived.

ᐋᓇ ᖁᔭᓐᓇᒦᖅᑐᖅ ᓃᑦ ᑯᓕᔅᑕᕗᕐᒧᑦ ᑖᓐᓇ ᐅᖅᑲᓕᒫᒐᖅ ᓴᖅᑭᒍᓐᓇᕐᓂᕋᖅᑐᓂᐅᒃ ᐅᑉᐱᕈᓯᓕᖅᑎᓚᐅᕐᒪᒍ ᐃᑲᔪᐃᓐᓇᖅᑐᓂᐅᑦᑐ ᓴᓇᔭᐅᓂᖓᓂ. ᐋᓇᐅᑉ ᖁᔭᒋᓪᓚᑦᑖᕐᒥᔭᖏᑦ ᖃᑕᙳᑎᖏᑦ ᐱᖃᓐᓇᕆᔭᖏᓪᓗ ᑲᔪᖕᒋᖅᓴᐃᑦᑎᐊᖕᒋᓐᓇᖅᐸᓚᐅᕐᒪᑕ ᐱᓇᓱᐊᕐᓂᓕᒫᖓᓂ.

Anna thanks Neil Christopher for convincing her that this project was possible and for helping her make it happen. Anna would also like to thank her family and friends for much-needed encouragement along the way.

ᓈᐱᑲ ᖁᔭᓐᓇᒦᖅᑐ ᐊᖓᔪᖅᑳᖏᓐᓂᑦ ᐃᓄᒃᑎᑐᑦ ᐅᖃᐅᓯᕐᒥᑦ ᖃᐅᔨᒪᑎᑕᐅᒐᒥ ᐊᒻᒪᓗ ᓄᓇᕋᐃᑦ ᒥᒃᓵᓄᑦ ᐃᓕᓐᓂᐊᖅᑎᑕᐅᖃᑦᑕᓚᐅᕋᒥ ᓱᕈᓯᐅᑎᓪᓗᒍ. ᖁᔭᓐᓇᒦᕐᒥᔭᖏᑦ ᐸᓂᖏᒃ, ᐱᖃᓐᓇᕆᔭᖏᑦ, ᐊᒻᒪᓗ ᐊᖏᔭᖏᑦ ᑲᔪᖕᒋᖅᓴᐃᑦᑎᐊᖅᑲᑦᑕᓚᐅᕐᓂᖏᑦ ᐱᓕᕆᓇᓯᑦᑎᓪᓗᒍ ᐊᔾᔩᙱᕈᓘᔭᖅᑐᓂᑦ.

Rebecca thanks her parents for keeping her Inuktitut strong and for teaching her about plants as a youth. She also thanks her children, friends, and sisters for supporting her through her many busy projects.

ᓴᓇᔾᐅᓂᖓ ᓴᖅᑭᑕᐅᓂᖓᓗ ᑖᒃᓯᒪ ᐅᖃᓕᒫᒐᐅᑉ ᐊᔪᙱᑎᑕᐅᓚᐅᖅᑐᖅ ᐃᓕᖅᑯᓯᓕᕆᔨᒃᑯᓐᓂᑦ ᐊᒻᒪᓗ ᑲᓇᑕᒥᑦ ᐃᓕᖅᑯᓯᑐᖅᑲᓕᕆᔨᒃᑯᓐᓂᑦ. ᐱᓕᕆᕙᑦᓕᐊᑎᑦᑐᑕᓗ ᑲᔪᙱᖅᓴᐃᖅᑲᑦᑕᔪᔪᓂᑦ, ᐃᓕᓴᕆᔪᒪᒻᒥᔭᕗᑦ: ᕇᑕ ᐊᖅᑭᐊᕈᖅ; ᔪᐋᓐ ᔅᑖᓴᓐ; ᐃᐊᓚᓐ ᓯᒡᓗᕐ; ᐹᑐᕆᒃ ᓕᑐ; ᓅᑦ ᒪᒃᑑᒪᑦ; ᐅᓘᑕ ᒫᑎᐅᓯ; ᓯᐊᓕ ᕉᔅ; ᓗᐃᔅ ᕝᓚᐦᐅᕐᑎ; ᑑᓂ ᕈᒦᑐ; ᑰᑎᔅ ᕈᐅᓚᓐ; ᐊᒻᒪᓗ ᓄᓇᕗᒻᒥᑦ ᐃᓕᓴᐃᔨᒃᓴᓄᑦ ᐃᓕᓐᓂᐊᕐᕕᒃ.

The creation and printing of this book was supported by the Department of Culture, Language, Elders, and Youth and by Canadian Heritage. For additional support during the project, we would also like to acknowledge: Rita Akearok; Joanne Stassen; Ellen Ziegler; Patrick Little; Noel McDermott; Oolootа Matiusi; Shelley Ross; Louise Flaherty; Tony Romito; Curtis Rowland; and, the Nunavut Teacher Education Program.

ᑖᓐᓇ ᐅᖃᓕᒫᒐᖅ ᑐᕌᖅᑎᑕᐅᔪᖅ ᐃᓕᒍᒪᔪᓄᑦ,ᐱᒻᒪᕆᐅᒍᓱᒃᑐᓄᑦ,ᓄᓇᕋᕐᓂᒃ ᐊᑐᖅᑕᐅᔾᔪᓯᐅᔪᕕᓂᕐᓂᒃ ᑲᔪᓯᑎᑦᑎᔪᒪᔪᓄᑦ.

This book is dedicated to all those who want to learn, value, and continue traditional plant use.

ᐋᓚᓯ ᔫᒥ
Aalasi Joamie

ᐋᓇ ᓰᒡᓗᕐ
Anna Ziegler

ᕆᐱᑲ ᓴᐃᓐᓄ
Rebecca Hainnu

ᐱᓕᕆᐊᑦ ᒥᒃᓵᓄᑦ

About this Project

ᖃᓄᕐᓕ ᖃᐅᔨᒪᑦᑎᐊᕐᓂᐊᕋᑦᑕ ᑭᓇᐅᓂᑦᑎᓐᓂᑦ,ᐋᓚᓯ ᐊᐱᕆᔪᖅ,ᓇᒥᒥᐅᑕᐅᓂᑦᑎᓐᓂᑦ ᖃᐅᔨᒪᙱᒃᑯᑦᑕ? ᐊᓈᓇᒥᓂᑦ ᐃᓕᓐᓂᐊᖅᐸᓪᓕᐊᓪᓗᓂ, ᐋᓚᓯ ᑕᑯᓐᓇᖃᑦᑕᓚᐅᖅᑐᖅ ᐃᓕᒃᑯᐊᓕᐅᖅᐸᑦᑐᓂᓗ ᓄᓇᕋᕐᓂᑦ ᓂᕚᖅᓯᐊᖑᓪᓗᓂ ᐸᓐᓂᖅᑑᒥᑦ 1940ᖏᓐᓂᑦ ᐊᒻᒪ ᑭᖑᓂᐊᒍᑦ ᐊᓈᓇᐅᓕᖅᑐᓂ ᓂᐊᖁᙳᒥᑦ 1960ᓂᑦ ᐅᓪᓗᒥᒧᑦ. ᐅᖃᓕᒫᒐᕐᒥᑦ ᑕᐤᕙᓂ ᓴᖅᑭᑕᐅᕗᑦ 18 ᓄᓇᕋᐃᑦ ᒥᒃᓵᓄᑦ, ᐋᓚᓯᐅᑉ ᖃᐅᔨᒪᔭᖏᑦ ᐃᖅᑲᐅᒪᔭᖏᓪᓗ ᓇᓂᔭᐅᒡᒍᔪᑦ ᐸᓐᓂᖅᑑᒥᑦ, ᓂᐊᖁᙳᕐᒥᑦ, ᓄᓇᕗᒻᒥᓗ.

How can we know who we are, Aalasi Joamie asks, *if we don't know about where we live?* Having learned from her mother, Aalasi observed and harvested plants as a little girl in Pangnirtung in the 1940s and later as a mother in Niaqunnguuq (Apex) from the 1960s to today. In this introductory guide to traditional plant use, Aalasi shares her life learning and memories of plants commonly found around Pangnirtung, Niaqunnguuq, and across Nunavut.

ᐋᓚᓯᒃᑯᒃ ᔫᐋᓂᒃᑯᒃ ᐃᖃᓗᐃᑦ ᑰᖓᑕ ᖃᓂᒋᔮᓂ ᖃᐅᔨᓴᖅᐸᓪᓕᐊᓪᓗᑎᒃ ᑖᒃᓱᒥᖓ ᐅᖃᓕᒫᒐᕐᒥᒃ.

Aalasi and Joannie Joamie near Iqaluit Kuunga (the Sylvia Grinnell River) during the research for this book.

ᐅᖃᓕᒫᒐᖅ ᐃᓗᓕᖃᖅᐳᖅ ᐱᒋᐊᕈᑎ ᐋᓚᓯᒦᙶᖅᑐᑦ, ᓇᓗᓇᐃᒃᑯᑕᑦ ᓄᓇᖃᐅᑉ ᐊᑑᑎᖏᓐᓂᑦ ᐊᑎᖏᓐᓄᓗ ᓇᓂᔭᕐᓂᕐᓂᐊᕐᒪᑕ, 18 ᐊᒻᒍᖅᓯᒪᔪᑦ ᓄᓇᖃᐃᑦ ᒥᒃᓵᓄᑦ, ᐊᒻᒪᓗ ᑎᑎᖅᓯᒪᔪᑦ ᐊᑐᒃᑲᓂᖃᒃᓴᐃᑦ ᓄᓇᖃᐃᑦ ᒥᒃᓵᓅᖓᔪᑦ.

This book includes an introduction by Aalasi Joamie, a summary of the uses and names of plants for easy reference, 18 sections on specific plants, and a list of resources for more about information about the plants in this book.

ᐅᖃᓕᒫᒐᐅᑉ ᐃᓗᓕᒃᓴᖏᑦ ᑲᑎᖅᓱᖅᑕᐅᓂᑯᐃᑦ ᐅᐱᕐᖔᒃᑯᑦ ᐱᓱᖃᔾᔪᑎᑦ ᐃᖃᓗᐃᑦ ᑰᖓᓂ (ᓯᐅᕝᕙᐊ ᒍᕆᓂᐅᓪ) ᐊᒻᒪᓗ ᓂᐊᖁᙳᒥᑦ ᐃᓚᓯᒪᓪᓗᑎᑦ ᓂᐱᓕᐅᖅᓯᒪᔪᓅᙶᖅᑐᓂ ᐋᓚᓯᐅᑉ ᐃᓪᓗᖓᓂ. ᓯᕗᓪᓕᖅᐹᑦ ᓂᐱᓕᐅᖅᓯᒪᔪᓂᑦ ᐋᓚᓯᐅᑉ ᐅᖃᓪᓚᒃᑕᖏᑦ ᓄᐊᑕᐅᓂᑯᑦ ᐋᓇ ᓯᒡᓗᒃᑯᓐᓂᑦ, ᐋᓚᓯᐅᓪᓗ ᐸᓂᖓ, ᔫᐋᓂ ᔫᒥ ᑐᓵᔨᐅᓪᓗᓂ, ᓂᐱᓕᐅᕈᑎᓕᕆᔨᐅᓪᓗᓂᓗ. ᑭᖑᓂᐊᒍ ᐃᓪᓘᑉ ᐃᓗᐊᓂ ᓂᐱᓕᐅᒃᑲᓂᕆᓪᓗᓂ. ᐊᐱᖅᓱᖅᑕᑦᑕᔪᑦ ᐋᓇ ᓯᒡᓗᖅ ᐊᒻᒪᓗ ᕆᑕ ᐊᕿᐊᕈᖅ ᐃᑲᔪᖅᑎᖃᖅᑐᑎᑦ ᔪᐋᓐ ᓯᑕᓴᓐ. ᐅᐱᕐᖓᒃᑲᓂᖓᑦ, ᐋᓇ ᓯᒡᓗᒃᑯᑦ ᕆᐱᑲ Hᐊᐃᓐᓄᒃᑯᑦ ᐱᓇᓱᐊᕈᓯᓚᒫᖅᒥᑦ ᐊᐱᖅᓱᖃᓂᒃᑕᒃᑐᑎᑦ ᐋᓚᓯᒥ ᐊᒻᒪ ᓂᕈᐊᖅᑐᑎᒃ ᓇᓚᐊ ᓄᓇᖃᐃᑦ ᑎᑎᖃᐅᑕᐅᓂᐊᕐᒪᖔᑕ ᓇᓗᓇᐃᔭᐃᓪᓗᑎᒃ, ᑐᑭᓯᒋᐊᕈᑕᐅᒃᑲᓂᖅᑐᑎᓪᓗ, ᓂᐱᓕᐅᖅᑕᐅᒃᑲᓂᖅᑐᓂᓗ ᐅᖃᓪᓚᒃᑎᓪᓗᒍ ᐋᓚᓯ ᐃᓅᓯᖓᑕ ᒥᒃᓵᓄᑦ. ᑎᑎᖅᑕᐅᓂᒃ ᐅᖃᓕᒫᒐᖅ ᐋᓇ ᓯᒡᓗᖅᒍᑦ ᐃᓄᒃᑎᑑᓕᖅᑎᑕᐅᓪᓗᓂ ᕆᐱᑲ Hᐊᐃᓐᓄᒻᒧᑦ.

The information in this book was gathered on walks around Iqaluit Kuunga (the Sylvia Grinnell River) and Niaqunnguuq and through additional interviews in Aalasi's house. The first research interviews were conducted on walks with Aalasi by Anna Ziegler and Aalasi's daughter Joannie Joamie, who interpreted from Inuktitut to English and helped with the recording equipment. Later in the season, we conducted several interviews indoors. These interviews were conducted by Anna Ziegler and Rita Akearok with assistance from Joanne Stassen. The following summer, Anna Ziegler and Rebecca Hainnu conducted a week of intensive interviews with Aalasi to select plants to be included in the book, to gather more information, and to record additional commentary by Aalasi about her life experiences. The plant sections were written by Anna Ziegler, based entirely on interview notes and transcriptions, and translated by Rebecca Hainnu.

ᐅᖃᓕᒫᒐᖅ ᐃᓗᓕᓕᒃ ᐋᓚᓯᐅᑉ ᖃᐅᔨᒪᔾᔪᑎᖓᓂᒃ, ᐃᓕᑕᕆᓂᖅᑎᖓᓂᓪᓗ ᐊᖓᔪᖅᑳᒥᓂᑦ ᐊᓯᖏᓐᓂᓗ ᐃᓄᖕᓂᑦ ᐸᓐᓂᖅᑑᖅᒥᐅᓂᑦ ᐱᕈᖅᓴᑎᓪᓗᒍ. ᐃᓄᐃᑦ ᐊᓯᖏᑦ ᖃᐅᔨᒪᔪᖅᑲᒃᑲᓐᓂᖅᑑᒐᓗᐊᑦ ᐊᓯᖏᓐᓂᑦ ᐊᔾᔨᒌᖏᑦᑕᖏᓐᓂᓗ. ᐊᒻᒪᓗ ᓄᓇᕗᐃᑦ ᐊᑎᖃᕐᑐᐊᓘᓪᓗᑎᑦ ᐊᔾᔩᖏᑦᑐᓂᑦ. ᓄᓇᕗᐃᑦ ᒦᖅᓱᓕᒫᕗᒃ ᑎᑎᕋᕈᖅᓯᒪᖏᑦᑐᒍᑦ, ᑭᓯᐊᓂᓕ ᐊᑕᐅᓯᕐᒧᑦ ᐃᓄᒻᒧᑦ ᖃᐅᔨᒪᔭᐅᔪᓂᑦ ᐊᑐᖅᑕᐅᓯᒪᔪᓂᑦ ᐸᓐᓂᖅᑑᒥᑦ ᓂᐊᖁᙳᐆᒧᑦ ᓇᓗᓇᐃᔭᐃᓇᕈᖅᓯᒪᕙᒍᑦ. ᓴᖅᑭᖁᔪᒥᓇᖅᐳᑦ ᐃᓐᓇᐃᑦ ᐊᓯᖏᑦ ᓇᑭᑐᐃᓐᓈᑐᑦ ᖃᐅᔨᒪᔭᒥᓂᑦᑕᐅᖅ ᑎᑎᖅᓯᒪᓗᒋᑦ ᓴᖅᑭᓗᑎᑦ ᒪᒃᑯᓐᓂᖅᓴᓄᑦ ᐊᑐᐃᓐᓇᐅᓂᖅᓴᐅᓂᐊᕐᒪᑕ ᓯᕗᓂᒃᓴᑦᑎᓐᓂᑦ.

The information in this book is a representation of Aalasi's personal knowledge, learned from her parents and other people she knew in Pangnirtung when she was growing up. Other people may know more or different things about the plants in this book. Also, there are many other names for the plants in this book. Our intention is not to present an objective body of knowledge about Arctic plants, but to record and share the subjective knowledge and experiences of one person from Pangnirtung and Niaqunnguuq. We hope that other Elders and knowledgeable community members will have opportunities to record and share their regional plant knowledge and memories so that a vibrant dialogue of Inuit traditional plant knowledge will be accessible to future generations.

ᖃᐅᔨᒪᒋᐊᕐᓂᐊᖅᐳᑎᑦ, ᓄᓇᕗᐃᑦ ᐊᒥᓱᑦ ᓴᙱᔫᒪᓇᐅᒪᑕ. ᐊᑐᕐᓂᐊᕐᓗᒋᑦ ᐅᔾᔮᖅᓱᑎᐊᓇᐊᖅᑲᕐᓂᐊᕐᕕᒋ! ᑕᐃᒪᑎᐊᖅ ᑎᑎᕋᕐᓱᖅᓯᒪᔪᒍ ᐅᖃᓕᒫᒐᕐᒥᑦ ᑕᕝᕙᓂ ᑭᓯᐊᓂ ᐸᓯᔭᒃᓴᐅᔾᔫᖏᑕᒎᑦ ᐊᑐᑎᐊᖅᑕᐅᖏᑦᑐᑎᑦ ᐊᑲᐅᙴᓐᓃᕈᑕᐅᑉᐸᑕ ᓄᓇᕗᐃᑦ. ᐃᓐᓇᑐᖃᕐᓂᑦ ᐊᐱᖅᓱᒃᑲᓐᓂᖅᑲᑦᑕᓇᑦ ᓄᓇᒃᓱᓐᓃᑦᑐᓂᑦ ᐊᐱᖅᑯᑎᒃᓴᖅᑲᒃᑲᓐᓂᕈᕕᑦ ᓄᓇᕗᐃᑦ ᒦᖅᓱᓄᑦ.

Please note, many plants have strong medicinal properties. Use them with care! We have done our best to provide accurate information in this book but we do not take responsibility for any adverse effects experienced as a result of using the plants. Please talk with Elders and other knowledgeable people in the community if you have questions about local plant uses and varieties.

ᑕᑯᑎᑦᑎᒋᐊᕈᑎ ᐋᓚᓯ ᔫᒥᒦᖕᒑᖅᑐᑦ
Introduction by Aalasi Joamie

ᑖᓇ ᑎᑎᕋᖅᓯᒪᔪᖅ ᐋᖅᑭᒋᐊᖅᑕᐅᓚᐅᖏᑦᑐᖅ ᐋᓚᓯᐅᑉ ᐅᖃᐅᓯᕕᓂᖏᑦ ᒪᓕᒐᒃᓯᒪᔫᒐᓗᐊᑦ. ᐋᓇᒃᑯᒃ ᕉᐱᒃᑯᒃ ᐊᐱᖅᓱᓚᐅᖅᓯᒪᒻᒪᑏᒃ ᖃᓄᖅ ᓄᓇᕋᐃᑦ ᒥᒃᓵᓄᑦ ᐃᓕᓚᐅᖅᓯᒪᒻᒪᖔ, ᖃᓄᖅ ᐃᓅᓯᖓᓄᑦ ᐊᑐᐃᓂᖅᑲᖅᓯᒪᒻᒪᖔᑕ, ᖃᓄᖅ ᐃᓱᒪᒻᒪᖔ ᑭᓱᐃᓐᓇᐃᑦ ᐊᓯᔾᔨᖅᐸᓪᓕᐊᓂᖏᑦ ᑕᑯᓯᒪᔭᖏᑦ ᐊᑐᖅᓯᒪᔭᖏᓪᓗ ᐃᓅᓯᖓᓂᑦ. ᐊᖏᕐᕋᖓᓂ ᔪᓚᐃ 2007ᖑᑎᓪᓗᒍ ᑎᑎᕋᖅᑕᐅᓂᑯᐃᑦ. ᕉᐱᑲᐅᑉᓗ ᓂᐱᓕᐅᖅᑕᐅᓯᒪᔪᑦ ᑎᑎᕋᖅᑐᓂᒋ ᖃᓪᓗᓈᑎᑑᓕᖅᑎᑦᑐᓂᒋᓪᓗ.

This introduction is an unedited transcription of oral commentary by Aalasi. Anna and Rebecca asked Aalasi to talk about how she learned about plants, what role plants have in her life, and how she feels about changes she has witnessed and experienced in her lifetime. It was recorded in July 2007 in her home. Rebecca transcribed the recording and translated it into English.

ᐊᓈᓇᒐ ᒪᑯᓂᖕᒐ ᓄᓇᔭᓂᑦ ᐅᓂᒃᑳᖅᑎᓪᓗᒍ,ᓂᓪᓕᐊᖅᑕᑦᑕᔪᖏᑦᑐᖕᒐ ᓈᓚᑐᐃᓐᓇᖅᑐᒍ ᐃᒫ ᐱᐅᔫᒻᒪᑕ ᒪᑯᐊ ᓄᓇᕋᐃᑦ ᐃᒪᐃᓕᐅᖅᑕᑦᑕᓚᐅᖅᓯᒪᒻᒪᑦ. ᓄᓇᕋᔭᓂᑦ ᓄᐊᑦᑎᑎᓪᓗᒍ ᖁᕕᐊᒋᓗᐊᖅᑕᑦᑕᓚᐅᖅᓯᒪᔭᕋ ᐃᓕᒐᓱᒃᑐᖓᓗ ᐊᓐᓂᐊᖅᓱᐃᑎᐅᓂᕋᐃᑦᑐᓂ ᐊᒻᒪᓗ ᓂᕿᐅᓂᕋᐃᑦᑐᓂ ᐊᒻᒪ ᓂᕿᐅᖏᓐᓂᕋᐃᑦᑐᓂ, ᓄᓇᕐᓂ.

ᐋᓚᓯ ᔫᒥ ᔪᓚᐃᒥᑦ, 2006, ᐃᖃᓗᐃᑦ ᑰᖕᒐᓂᑦ (ᓯᐅᓪᕕᐊ ᒍᕆᓂᐅᓪ).

Aalasi Joamie in July, 2006, near Iqaluit Kuunga (the Sylvia Grinnell River).

My mother taught me which plants were good. I would listen as she talked about them. I would watch my mother collecting plants. Some plants are edible and some are not. Others have medicinal uses.

ᓄᓇᕋᐃᑦ ᓂᖅᑭᐅᓂᕋᖅᑕᖕᒋᑦ ᓇᓗᓇᐃᕋᓯᑦᑕᕆᖅᑲᑦᑕᓚᐅᖅᓯᒪᔭᒃᑲ ᑎᑎᖅᑲᑎᒨᖕᒋᑦᑐᕐᓚ, ᑭᓱᐊᓂ ᐃᒃᑯᐊ ᐊᕙᓃᔪᔪᖅ ᓂᖅᑮᑦ ᐃᒫ ᑕᐅᑐᒐᓯᐊᖅᑐᖓ. ᐊᒻᒪᓗ ᑎᒍᒥᐊᖕᒋᓐᓇᖅᑲᑦᑕᔪᔪᖕᒐ ᐅᕙᖓ ᐱᐅᒋᔭᖕᒋᓐᓂᑦ ᓄᓇᕋᕐᓂᑦ. ᑕᐃᒃᑯᐊ ᑎᒍᒥᐊᖅᑕᒃᑲ ᓄᓇᕋᐃᑦ ᐱᕈᖅᑎᒐᓯᐊᓯᑦᑕᖅᑲᒃᑲ. ᐅᑭᐅᑭᑦᑐᕈᓘᑦᓗᖓ, ᒫᓂᖅᑲᐃ ᐅᑭᐅᖅᑲᖅᑐᖓ 10, 11ᖑᔪᒃᓴᐅᓐᓂᖅᐸᖓ. ᐱᕈᖅᑎᒐᓯᐊᖅᑲᑦᑕᓚᐅᖅᓯᒪᔭᒃᑲ, ᖃᑦᑖᕐᔪᒥᓃᑦ ᖁᖑᔪᒃᓯᒪᔪᖅ ᓯᓚᒥᑦ ᓄᐊᑦᑐᒋᑦ. ᖃᖕᒑᓗᒃᑯᑦ ᖃᑦᑖᕐᔪᕕᓂᕐᒥᑦ ᑕᑯᑦᑕᓯᓐᓇᓕᖅᑲᑦᑕᖅᑐᖓ, ᖃᓪᓗᓈᖅᑲᖕᒋᑦᑐᒦᖅᑲᑦᑕᖅᑐᑕᓕ.

I would make mental notes about which plants were edible. I would try to remember places where they were found. I would always have plants that I liked with me. I was around ten or eleven years old when I first began to experiment with cultivating plants. I would use old tin cans for potting them. We lived in an area where there were no Qallunaat, so I only found tin cans once in a long time.

ᐊᓈᓇᒐ ᕿᓯᖕᒥᑦ ᕿᑦᓯᓇᑎᓪᓗᒍ ᓯᓚᒥᑦ ᑕᐃᒪᐃᑦᑐᓂᒃ ᓄᐊᑦᑎᖅᑲᑦᑕᓯᑦᑕᖅᐸᖓ ᐱᕈᖅᑐᓂᑦ ᑕᖅᑳᙵ ᓯᓚᒥᑦ ᐱᐅᒋᔭᖕᒋᓐᓂᑦ ᐅᕙᖓ. ᐊᓯᐃᑦᓛ ᖃᐅᑉᐸᕈᕋᐃᒻᒪ ᐅᑕᖅᑮᓐᓇᖅᑕᖅᑯᖓ ᐱᕈᕐᒪᖔᑕ ᑕᑯᕆᐊᖕᒋᓐᓇᖅᑲᒃ. ᐱᕈᖅᑕᖅᑯ ᐊᓈᓇᓐᓄᑦ ᐅᖃᑦᑕᖅᐸᖓ ᐱᕈᕐᒪᖔᑕ ᐃᒫ ᐊᓈᓇᒐ ᐃᒪᐃᓕᑦᑕᖅᑯ, “ᐱᐅᒋᒍᕕᒋᑦ ᑭᓱᐊᓂ ᐱᕈᓚᖕᒐᔪᐃᑦ, ᖁᕕᐊᒋᖕᒋᒃᑯᕕᒋᑦ ᐱᕈᔾᔫᖕᒋᑦᑐᐃᑦ. ᖁᕕᐊᒋᒍᕕᒋᑦ ᑭᓱᐊᓂ”. ᑕᐃᒪ ᐱᕈᖅᐸᓪᓕᐊᖃᑦᑕᕐᒪᑕ.

My mother often prepared skins while I planted flowers. She would always say to me, if you appreciate them, they will grow. If you like what you are doing, they will grow well.

ᐃᓱᒪᖃᑦᑕᓯᑦᑕᕆᕚᖓ ᒫᙵ ᓄᓇᑦᑎᓐᓂᑦ ᐱᐅᒋᔭᖕᒋᓐᓂᑦ ᐅᕙᖓ ᐱᕈᖅᓯᐊᓂᑦ ᐋᖅᑭᒃᓱᐃᓗᖓ ᖃᐅᓯᖅᑐᒍᑦ ᓄᐊᑦᓗᒋᑦ. ᐃᒫ ᕿᑲᖅᓯᔪᓐᓇᑦᑎᐊᖃᑦᑕᔪᖕᒋᓐᓇᒪ, ᐱᕈᖅᓯᐊᓕᓅᓐᓇᖅᑑᔭᖅᑲᑦᑕᖅᑐᖓ ᖃᐅᔨᕙᑦᑕᐊᓯᒪᒐᒪ. ᐃᓚᖕᒋᑦᓗ ᓂᕆᕙᑦᑐᒋᑦ. ᐅᖅᔫᒋᐊᖅᑐᖅᐸᑦᑐᖕᒐᓗ ᐱᕈᖅᑐᓂᑦ, ᐊᒻᒪᓗ ᑕᐃᒃᑯᐊ ᓂᕿᐅᓂᖕᒋᑦ ᓇᓗᓕᕐᔭᐃᖅᓯᒪᓪᓗᒋᑦ. ᐊᓂᑯᓗᓐᓄᓗ ᒥᓇᖅᐸᑦᑐᖓ ᐋᒪᕐᓂᑦ ᖁᐊᕋᕐᓂᑦ, ᐸᐅᓐᓇᓂᑦ. ᐊᑯᑦᑎᙳᐊᖅᑐᐊᓘᕙᑦᑐᖓᓗ ᐱᐅᓯᖕᒋᑦᑎᐊᖅᑎᓪᓗᒍ ᐃᓚᖕᒐᓂ ᓂᕆᕙᑦᑕᕋ, ᐃᓚᒃᑲ

ᓇᕐᕈᕙᑦᑐᑦ. ᐊᒻᒪᓗ ᑕᐃᒃᑯᐊᓗ ᐱᕈᖅᓯᐊᑯᓗᐃᑦ ᒪᓕᒃᑳᑯᓗᐃᑦ, ᑕᐃᒃᑯᐊ ᓄᐊᑦᑕᖅᑲᒃᑲ. ᐊᓯᐃᑦᓛ ᐱᐅᓯᕙᑦᓕᖏᓐᓇᕋᒥᑦ ᑕᐃᒫ ᐅᑕᖅᑭᑦᑎᐊᖅᓯᒪᒐᒪ ᐱᕈᖅᐸᑦᑕᐊᖏᓐᓇᖅᐸᑦᑐᑦ.

Then I began to plant flowers in the moist soil. I remember I was always busy with plants. I collected them to eat and I mixed them in oil. Sometimes I would collect edible plants for my brother. Sometimes I would make *alu* (pudding) and, before it was ready, I would eat it. Not all my family members were willing to try it.

ᑕᐃᒪᙵᓂᐊᓗ ᐅᑭᐅᑭᑦᑐᕈᓘᒐᒪ ᑕᐃᒫ ᐱᕈᖅᓯᐊᓂᑦ ᖃᐅᔨᒐᓯᐊᖅᓯᒪᒐᒪ ᐃᓛ ᐱᐅᒋᓯᒪᒐᒃᑮᑦ ᑕᒪᓐᓇ ᑕᕝᕙ ᐃᓅᓯᕆᓯᒪᔭᕋ. ᐅᑦᓗᒥᒧᑦ ᑎᑭᑦᑐᒍ ᓄᓇᕋᑕᓈᓐᓇᖅᓯᖓ. ᐃᓄᑦᑐᖓ ᐱᓱᒃᓯᒪᑦᓗᖓ ᐱᕈᖅᑐᐃᑦ ᐱᔾᔪᑎᒋᑦᓗᒋᑦ ᖁᔭᓐᓇᒦᖅᑐᖓ ᐊᒻᒪᓗ ᓂᕿᐅᓂᖏᑦ ᐅᐱᒋᑦᓗᒋᑦ.

I would keep *malikkaat* (mountain avens) for flowers. It wasn't long before they would bloom. Throughout my life, I have been interested in plants. I have always inquired about them. When I am walking by myself, I feel thankful for all the plants. I appreciate that they are food.

ᐊᕿᐊᑦᑐᒐᓚᓲᖑᒻᒪᕆᑉᐸᔪᔪᖓ ᐱᕈᖅᑐᓂ ᓂᕆᑦᓗᖓ ᖁᖑᓚᕐᓂᑦᓗ ᐊᖏᔪᐊᓗᓐᓂᑦ ᐊᔾᔭᖅᓯᓲᖑᔪᖓ ᐊᒻᒪᓗ ᒪᑯᐊ ᐊᒻᒫᖅ ᑎᒍᒥᐊᖅᑕᐅᓯᐅᑦ ᒥᓇᕆᕙᑦᑐᒥᑦ, ᐃᓛ ᐱᐊᕋᖅᑲᖏᒃᑲᓗᐊᕋᒪ, ᐱᐊᕋᑕᓐᓄᑦ ᑐᓂᕙᑦᑐᒥᑦ. ᑕᒪᒃᑯᐊ ᐊᔪᓇᖅᓲᔾᔭᐅᔾᔪᑎᒋᓯᒪᔭᒃᑲ ᐊᑐᕋᓱᐃᓐᓇᖅᑕᒃᑲ. ᐅᑦᓗᒥᑕ ᐃᓐᓇᐅᑕᖅᑐᖓ ᐱᕈᖅᓯᐊᖕᒑᓄᑦ ᔫᓯᒪᒐᒪ. ᖃᑦᑐᓈᕐᑕ ᐱᕈᖅᑐᕐᑯᑎᖏᓐᓄᑦ. ᑕᐃᒫ ᑭᓯᐊᓂᐅᔫᖅᑐᖅ ᐱᕈᖅᓯᐊᖅᑲᕐᑐᖓᐅᔫᖅᑐᖅ. ᐱᕈᖅᓯᐊᖅᑲᖏᒃᑯᒪ ᐱᔪᓐᓇᖅᑐᖅᔫᖏᑦᑐᖓ, ᐋᒃᑲᐅᒻᒥᔫᒐᓗᐊᖅ.

ᐋᓚᓯᐅᑉ ᐊᓈᓇᖓ. ᓇᓗᔭᐅᔪᑦ ᖃᖓ ᑭᐊᑦᑐ ᐊᔾᔨᓕᐅᕐᓂᕐᒪᖔᒍ. ᐊᔾᔨᙳᐊᖅ ᑐᓂᔭᐅᔪᖅ ᐋᓚᓯ ᔫᒥᒧᑦ.

Aalasi's mother. Date and photographer unknown. Photograph provided by Aalasi Joamie.

Sometimes I could get a good fill from eating plants, especially from eating *qunguliit* (mountain sorrel). I was able to collect them and carry quite a lot. Even before I had my own children, I would gather roots and give them to infants to suck on. I tried to use them as I had been taught.

ᐊᑦᓯᐊᓗᒃ ᐱᐅᕆᔭᒃᑲ, ᐊᑦᓯᐊᓗᒃ ᐅᕙᓐᓅᖓᔪᐃᑦ. ᐃᓛ ᐃᓪᓗᒥᐅᖅᑲᑎᒥᔫᔾᖅᑲᑕᒃᑲ, ᓂᑦᑕᐊᖅᑲᑎᒥᔫᔾᖅᑲᑦᑕᖅᑕᒃᑲ. ᑕᕝᕙ ᓱᕈᓯᐅᓪᓗᖓ ᓂᕕᐊᖅᓯᐊᙳᓪᓗᖓ, ᐅᕕᒃᑲᐅᓪᓗᖓ ᐊᒻᒪ ᐅᕕᒃᑲᕈᖅᐸᑦᓚᐊᓪᓗᖓ, ᐊᒫᖅᑲᑦᑕᓚᖅᑐᖓᓗ ᕿᑐᙵᖅᑲᓚᖅᑐᖓ ᐱᕈᖅᓯᐊᓂᑦ ᐱᕕᔾᔭᐊᖅᑲᖅᓯᒪᖓᒪ. ᑕᐃᓛ ᕿᑐᙵᒃᑲ ᑕᓯᐅᖅᑐᒥᑦ ᐅᖅᑲᓚᒪᕕᒥᖅᑲᑦᑕᖅᓯᒪᔭᒃᑲ ᐱᕈᖅᑐᐃᑦ ᐱᐅᔫᒻᒪᖔᑕ ᐊᒻᒪᓗ ᐱᐅᙱᑦᑑᒻᒪᖔᑕ ᐱᕈᖅᑐᐃᑦ. ᑕᐃᒪᐃᒃᑑᖅᑲᑦᑕᖅᓯᒪᔪᖓ. ᐋᓐᓂᐊᓯᐅᑦᑎᐅᔪᓐᓇᖅᓂᖏᑦ ᖃᐅᔨᒪᔭᒃᑲ, ᐱᐅᔪᐃᑦ ᖃᐅᔨᒪᔭᒃᑲ, ᐊᒻᒪᓗ ᐋᓐᓂᖅᓇᖅᓂᖏᑦ ᖃᐅᔨᒪᔭᒃᑲ. ᑕᒻᒪᖅᑐᕕᓂᐅᒍᒪ ᐱᓯᑦᓗᖓ ᐅᔭᕋᐃᑦ, ᓄᓇᐃᑦᓗ, ᑕᒻᒪᓈᒃᑯᑎᒥᓗᒥᑦ ᐊᑖᑕᒪ ᐊᔪᓇᖅᓲᔾᔨᔾᔪᑎᒥᑦᓗᓂᒥᑦ, ᑕᐃᒪᐃᒐᓯᐊᖅᑲᑦᑕᖅᓯᒪᒻᒥᔪᖓ ᑕᕐᓇᓯᒪᔪᒍᑦ ᑕᒻᒪᖁᔭᕐᒪᖔᕐᒪ ᐃᓐᓈᕈᐃᑦ, ᐊᓄᓂ ᐊᒻᒪ ᐃᕕᒃᓯᒐᐃᑦ ᐊᑐᖅᑐᒥᑦ.

As an elder, I have turned to the flowers of the Qallunaat as well. It is as if I am unable to live without plants. I really like plants. They have been a part of my childhood, my adolescence, and my motherhood. I have taken my toddlers out on walks with me. I have tried to pass on my knowledge of plants to my children. I know which plants are edible, which are harmful, and which have medicinal uses. My father also taught me how to use plants as indicators, as a compass is used. By using rocks, the positions of plants, wind, and hills, you can find your way back. I have learned the use of these indicators through trial and error.

ᐃᔨᒧᑦ ᐊᑐᖅᑐᑦ ᐃᔨᓯᐅᑏᑦ ᐊᑐᕋᓱᖅᑲᑦᑕᖅᓯᒪᔭᒃᑲ. ᓲᕈᓗᓗᐊᖅᑐᖓᓗ ᒪᓃᑦ ᓂᕆᖅᑲᑦᑕᖅᓯᒪᔭᒃᑲ ᐋᖅᑭᒍᑎᒥᓇᓱᒃᑐᒥᑦ. ᓂᕐᓇᐃᓗ ᖃᐅᔨᒪᒻᒥᔭᒃᑲ ᐋᓐᓂᐊᓯᐅᑎᐅᔪᓐᓇᖅᑐᖅ. ᐊᓈᓇᒐ ᐅᕙᓐᓂᑦ ᖃᐅᔨᑎᑦᑎᖅᑲᑦᑕᖅᓯᒪᙱᑉᐸ ᖃᐅᔨᒪᒐᔭᙱᑕᒃᑲ. ᐅᕙᓐᓂᑦ ᐃᓕᓐᓂᐊᖅᑎᑦᑏᓐᓇᖅᑲᑦᑕᕐᒪ ᖃᐅᔨᒪᓕᖅᑭᔭᒃᑲ ᐅᓂᒃᑳᙱᓐᓇᖅᑲᑦᑕᕐᒪ ᐱᓯᑐᓄᒃ. ᒪᕿᐃᓐᓇᐅᙱᒃᑲᓗᐊᖅᑐᓄ ᐃᓛᓐᓂᓗ ᒪᕿᐃᓐᓇᐅᓪᓗᓄ.

I have used *ijisiuti* (river algae) to relieve a sore eye. I have used *maniq* (lamp moss) to relieve heartburn. I know that *nirnait* (snow lichen) has medicinal uses. My mother taught me these things. If she had not shown me these things, I would not know them.

Sometimes we would walk by ourselves and other times we had company, but I always paid attention.

ᐊᒻᒪᓗ ᐃᖅᑲᓗᐃᑦ ᑰᒥᒡ ᑕᑯᕋᓈᖅᑐᕆᑦ ᖃᐅᔨᑎᑦᑎᖃᑦᑕᖅᓯᒪᔪᖅ ᐊᓈᓇᒐ ᐅᕙᓐᓂᑦ. ᓇᒧᖖᒐᓲᖕᖑᒻᒪᖖᒑᑕ ᐱᕈᕚᖕᒐᑕ ᐃᖅᑲᓗᕋᑳᑯᓗᐃᑦ, ᐊᒻᒪᓗ ᓱᓲᖕᖑᒻᒪᖕᒑᑕ. ᖃᓄᑎᒫᑦᑎᐊᖅ ᐊᓈᓇᒪ ᐊᔪᓇᖅᓲᑎᓯᒪᒌᖕᒐ.

My mother taught me about many things. From her, I learned where the char fry go when they grow up and what they do. My mother taught me about everything.

ᐱᕈᖅᓴᓯᒪᒐᒪ ᐸᓂᖅᑑᒥ. ᐅᑭᐅᑭᑦᑐᕈᓘᑦᑐᖕᒐ ᐸᓂᖅᑑᔫᖅᑐᑕ. ᐱᕈᖅᓴᓯᒪᔪᖕᒐ ᐸᓂᖅᑑᒥᑦ, ᑭᓯᐊᓂᓕ ᖁᑐᖖᒐᖅᑲᖅᓕᖅᑐᖕᒐ ᐱᖕᒐᓱᓂᑦ ᒪᐅᖖᒐᓕᓚᐅᖅᓯᒪᔪᖑ (ᓂᐊᖁᖖᖑᒧᑦ) ᐅᑭᐊᒃᓵᖑᓗᒃᑯᑦ, ᐃᒃᑮᖅᓇᖅᓯᔪᐊᓘᑎᓪᓗᒍ. ᑭᓯᐊᓂ ᐊᔾᔨᒋᓚᐅᖅᓯᒪᖖᒋᑕᖕᒐ ᒪᐅᖖᒐᖁᑦ. ᒪᓐᓇ ᐊᔾᔨᒌᖕᒋᔾᔪᑎᒋᓚᐅᖅᓯᒪᔫᖕᒐ ᑕᑉᐹᖖᒑᖁᑦ ᐱᑦᑎᐊᖅᑖᓘᓪᓗᑕ ᖃᓄᓕᒫᖅ ᐊᐃᑦᑐᖅᑕᐅᕙᑦᑐᑕ.

We moved to Pangnirtung when I was very young. It was not until I had three children that we moved here [to Niaqunnguuq (Apex)]. We moved here in the fall, when the weather was becoming very cold. It was very different here.

ᑐᖖᒐᓯᐃᓐᓇᖅᑐᐊᓘᓪᓗᖕᒐ ᐃᓕᖁᓯᖕᒋᑦᑎᐊᖅᑐᖕᒐ ᐊᓯᔾᔨᖅᑐᐊᓘᑳᓪᓚᒃᑐᖕᒐ, ᖁᕝᕙᐊᓇᖅᑐᐃᓐᓇᐅᔫᖅᑐᑎᑦ. ᑭᓯᐊᓂᓗ ᐃᓚᒃᑲ ᕿᒪᒍᒪᖕᒋᒃᑲᓗᐊᖅᑐᕆᑦ ᐊᓈᓇᒃᑯᒃᑲ. ᓱᓇᐅᕝᕙ ᑕᑯᒃᑲᓂᔾᔪᓐᓃᖅᑐᕆᑦ ᐊᓈᓇᒃᑯᒃᑲ. ᐊᓯᔾᔮᒃᑲᓪᓚᑦᑐᐊᓗᔾᔭᓚᐅᖅᓯᒪᔪᖅ. ᖃᐊᖅᑲᑦᑕᓚᐅᖅᓯᒪᒐᒪ ᐃᓚᓐᓅᕈᒪᓪᓗᖕᒐ, ᑭᓯᐊᓂ ᑐᖖᒐᓯᓚᖕᒐᔪᐊᓘᓪᓗᑕ.

People were kind and willing to help us out through gifts. I was no longer afraid. I then had confidence. Everything was exciting. However, there were times when I cried, longing for my family. I never saw my mother again.

ᑕᒪᓐᓇᓗ ᐃᓕᓯᒪᓇᒍ ᓄᓇ, ᐃᓄᖕᒋᑦᓗ ᐃᓕᓯᒪᓇᒋᑦ. ᓱᓇᐅᕝᕙ ᐊᖕᒋᖅᑯᕈᒪᔾᔮᒍᓐᓂᖅᑐᖕᒐ, ᑕᑉᐸᐅᖖᒐᕈᒪᔾᔮᖕᒋᑦᑐᖕᒐ. ᑕᐃᒪ ᐊᒃᓱᕈᖅᓇᖅᑐᒃᑰᖅᐸᓪᓕᓚᐅᖅᓯᒪᔪᖕᒐ ᐃᓐᓇᕈᖅᐸᓪᓕᐊᓕᖅᑐᖕᒐ ᕿᑐᖖᒐᒃᑲᓗ ᐊᒥᓱᕈᖅᐸᓪᓕᐊᓕᖅᑐᑎᒃ. ᐃᓚᒃᑲᓗ ᓇᔾᔭᖅᑎᒋᓇᒋᑦ. ᑭᓯᐊᓂ ᐊᔾᔨᖅᓇᖅᑐᒃᑯᕋᓗᐊᖅᑕ ᐱᔪᓐᓇᖅᑕᒃᑯᖕᒋᓐᓇᖅᑎᑕᐅᖅᑲᑦᑕᓚᐅᖅᓯᒪᔪᖑ.

We were very well received here. I was in an unfamiliar land with unfamiliar people. But in the end, I would never long to go back up there [to Pangnirtung]. I was never homesick again. When I got older and had more children, I went through hardships. I had to persevere without the support of my extended family because I did not live close to them. Fortunately, we were able to endure those hardships.

ᑖᒃᑯᐊ ᕿᑐᕐᖓᒃᑲ ᐱᕈᖅᐸᓪᓕᐊᓛᖅᑎᓪᓗᒋᑦ ᐅᖃᐅᑎᖃᑦᑕᓚᐅᖅᓯᒪᔭᒃᑲ ᐱᕈᖅᑐᐃᑦ ᖃᓄᖅ ᐊᑐᖅᑕᐅᓯᐅᖑᒻᒪᖔᑕ ᐊᒻᒪᓗ ᐊᐅᓪᓚᖅᓯᒪᓗᑎᑦ ᓂᕿᒃᓴᐃᒍᓐᓂᕋᓗᐊᕈᑎᑦ ᐱᕈᖅᑐᓂᑦ ᓂᕿᒃᓴᖃᕈᓐᓇᕋᓯᐊᖅᑯᓪᓗᒋᑦ. ᑕᐃᒫ ᐅᖃᐅᑎᖃᑦᑕᓚᐅᖅᓯᒪᔭᒃᑲ. ᐊᐅᓪᓚᖅᓯᒪᓪᓗᑕ ᐊᔪᕆᖅᓰᑎᓇᓯᐊᖅᐸᑦᑐᒋᑦ ᓂᕿᐅᓂᖏᓐᓂᑦ ᐱᕈᖅᑐᐃᑦ, ᓂᕿᐅᖏᓐᓂᖏᓐᓂᑦ ᐃᓚᖏᓐᓂᑦ.

As my children were growing up, I taught them about plants. During camping trips, I showed them which plants were edible and which ones were not. I wanted them to know that if they ever got stranded on the land, they could draw upon many resources available to them.

ᑕᐃᒪ ᐅᓪᓗᒥᐅᓕᖅᑐᑦ ᐅᕙᒃᑲᐃᑦ ᖃᐅᔨᒪᖏᑦᑎᐊᓕᕐᒪᑕ, ᐅᕙᒃᑲᐅᖏᑦᑑᒐᓗᐊ ᑕᒪᓐᓇ ᓄᓇᔭ

ᒥᕐᖑᐃᖅᓯᕐᕕᒻᒥᑦ ᓂᐊᖁᙳᒥᑦ (ᐊᐃᐸᒃᔅ), ᑕᒫᓂᑦ ᐋᓚᓯ ᐃᓕᒃᑯᐊᓕᐅᓕᖅᑐᖅ ᑕᐃᒪᙵ ᓅᓚᐅᖅᓯᒪᒐᒥᑦ ᐸᓐᓂᖅᑑᒥᑦ 1960ᖏᓐᓂᑦ.

Coastal park in Niaqunnguuq (Apex), where Aalasi has harvested plants since she moved from Pangnirtung in the 1960s.

ᓂᕐᑭᒃᓴᖅᑕᕐᑲᕐᒪᖔᑦ ᓄᓇᒍᑦ ᐱᕈᖅᑐᓂᑦ. ᐱᑕᓕᐊᓘᑎᓪᓗᒍ, ᐅᔾᔨᕆᑦᑎᐊᕋᓱᐊᕐᓗᒋᑦ ᑲᒪᒋᓗᒋᑦ ᑕᒪᓐᓇ ᓂᕆᔭᒃᓴᖅᑕᓕᐊᓗ ᓄᓇᒍᑦ. ᐃᒫ ᓂᕆᔭᒃᓴᖅᑕᓕᐊᓘᔾᔮᙱᑦᑑᒐᓗᐊᖅ ᓂᕐᑭᓪᓚᕆᐊᓗᓐᓂᑦ. ᑭᓱᐊᓂ ᓂᕆᔭᒃᓴᖅᑕᕐᑲᕐᒪᑦ ᐃᒫ ᐆᓪᓗᒋᑦ ᐊᒻᒪ ᓂᕆᑐᐃᓐᓇᕋᓗᐊᕐᓗᒋᑦ. ᐱᐅᓪᓗᓂᓗ ᓄᓇ ᐊᒻᒪ ᕿᐅᓪᓗᓂ ᐅᖅᑰᔾᔪᑎᒃᓴᖅᑕᓕ. ᐊᔪᕆᖅᓲᑎᒋᐊᒃᑲᓐᓂᖅᐸᒃ ᐅᕝᕙᒃᑲᐃᑦ. ᑕᒻᒪᖅᓯᒪᓕᕐᓂᕈᒃᓯᑦ ᓇᒦᓪᓗᓯ, ᕿᔪᒃᑕᖅᓯᒪᓗᓯ ᓄᐊᑦᑎᓯᒪᓗᓯ. ᐊᒻᒪᓗ ᐃᓐᓈᕈᔾᔫᒃ ᓴᓂᐊᓂ ᐅᖅᑯᐊᓕᓂᕐᒥᑦ ᕿᓂᕐᓗᓯ ᐊᑐᖅᑐᐃᑦ. ᐅᕝᕙᒃᑯᐅᔪᓯ ᐅᕐᑯᐅᑎᒋᐊᖅᐸᒃᓯ. ᐊᓄᕆᒧᑦ ᓵᙲᙱᑎᓪᓗᒍ ᐅᖅᑯᐊᒧᑦ ᓵᙲᙳᕐᓗᑎ ᐊᒡᒍᕋᖅᑕᐃᓕᓗᓯ.

Now, young and old alike seem to be ignorant about tundra vegetation. They do not seem to realize that there is a lot of food out there. I know it must look like there is not much food, but there is. We must notice the plants and we must look after them well. There are foods out there that we can cook or eat after picking them. Plants are healthy for us. I am talking to the younger generation now. Plants can be used as survival tools. For example, if you get lost during a hike, you can pick *qijuktaat* (white heather) for a mattress. Go behind a rock or ditch facing away from the wind and use the qijuktaat for warmth.

ᐊᒻᒪ ᑰᒻᒥᑦ ᐃᑳᕈᓐᓇᐃᓕᐅᖅᑲᖅᑕᐃᓕᓗᓯ. ᓯᑕᓗᓰᕋᓗᒻᒪᑦ ᓯᑕᓗᑦᑐᐹᓗᑕᐅᖅᑎᓐᓇᒍ ᐃᑳᕆᐊᕐᑲᑦᑕᕐᓗᓯ. ᓯᑕᓗᕋᑖᖅᑎᓪᓗᒍ ᐃᑳᕈᓐᓇᐃᓕᓰᕋᓗᑲᑦᑕ. ᐅᒃᑯᐊᑦᑕᐅᖅ ᖃᐅᔨᓴᑦᑎᐊᕐᑲᑦᑕᕐᓗᒋᑦ ᑰᒃ, ᐊᒻᒪ ᑎᓂᓐᓂᖅ ᐊᒻᒪ ᐱᑐᕐᓂᐊᓗᒃ. ᑕᒪᒃᑯᐊ ᖃᐅᔨᓴᕆᐊᕐᑲᖅᑕᐃᓐᓇᕆᑲᑦᑎᒍ. ᐊᑏᓪᓗ ᐱᕈᖅᑐᓂ ᖃᐅᔨᔪᒪᒍᑦᓯ ᐊᐱᖅᓯᕈᓐᓇᕐᑲᑦᑕᕋᔭᖅᑐᓯ ᐅᕙᓐᓂᓪᓘᓐᓃᑦ ᑕᒪᒃᑯᐊ ᖃᐅᔨᒪᒐᒃᑭᑦ ᐱᕈᖅᑐᐃᑦ ᐊᒻᒪᓗ ᑭᓱᑐᐃᓐᓇᐃᑦ.

I would also like people to know that they should not be scared to cross rivers. They should try to cross rivers before the rain. Rivers can become very deep after it rains, so people should try to cross them as soon as they can. I think people should become more aware of the behaviour of tides, full moons, and rivers. If you want to learn more about plants and other things, ask people who know about them, like I do. I know about other things, too.

ᐃᓐᓇᐅᓕᖅᐸᑦᑕᐊᓕᖅᑎᓪᓗᙱᑲᓕ ᐊᑐᖅᓯᒪᒍᙱ ᑕᒪᒃᑯᓂᙱ. ᐱᕈᖅᐸᑦᑕᐊᓂᐊᖅᑐᓯ ᖃᐅᔨᕙᑦᑕᐊᒋᐊᕐᑲᕐᓂᐊᖅᑐᓯ ᖃᓄᕈᓘᔭ ᐱᕈᖅᑐᐃᑦ ᖃᓄᐃᑦᑑᒪᙱᑦ ᐊᒻᒪᓗ ᑕᒻᒪᖅᓯᒪᓕᖅᑎᓪᓗᑕ

ᓄᓇᒥᑦ ᓇᒍᑦ ᓵᖖᒐᒋᐊᖃᕐᒪᖕᒑᑦᑕ, ᓇᐅᒃᑯᓪᓗ ᒪᔪᓇᐊᖃᕐᒪᖕᒑᑦᑕ, ᐊᖅᑲᓇᐊᖃᕐᒪᖕᒐᑕᓗ. ᐅᓗᔾᔪᐆᓂᑦ, ᐅᓕᓐᓂᖅ ᖃᐅᔨᓴᑦᑎᐊᕐᓗᒋᑦ ᐊᒻᒪᓗ ᓄᓇᖓᕐᓂᑦ ᐱᕈᖅᑎᐊᓲᖑᓂᖕᒋ ᐊᒻᒪ ᐊᓐᓂᐊᓯᐅᑎᐅᓂᖕᒋᓐᓂᑦ. ᑕᐃᒫᒐᓪᓚᒃ.

Growing up, I was immersed in the knowledge of plants. You must take the time to learn about them. You should know which plant does what. They can help you find your way if you are lost on the tundra. They are indicators of direction. But, also, you can use them to identify areas where it is safe to climb up or down. You must also learn about the high tide, such as when and where it occurs. You should learn about the different uses of plants, which ones are edible and which ones are used for medicine. These are the sorts of things you should learn.

ᒪᑯᓂᖕᒐ ᐊᓯᔾᔨᖅᐸᓪᓕᐊᓕᖅᑐᓂᑦ ᐅᔾᔨᕈᓯᓗᐊᖖᒍᐊᖅᓯᒪᕗᖕᒐ. ᒪᐅᖖᒐᔪᒐᑦᑕ ᖃᕐᒪᖃᔪᒐᑦᑕ ᐅᐱᓐᓇᖓᓂ ᐃᑦᓗᓕᑦᓚᑎᐊᓗᑦᑕᖃᔪᖕᒋᒻᒪᓪᓕᑦᓕ. ᒪᐅᖖᒐᖓᑦᑕ ᐃᓗᓕᐅᒐᓗᐊᑦ ᐃᓚᐃᓐᓇᖕᒋᑦ. ᑕᐃᑲᓂ ᖃᕐᒪᒦᑦᑐᑦ ᐃᑦᓗᑖᑖᕐᓂᐊᕐᓂᖓᑦᑕᐅᓕᖓᑦᑕ ᐃᓯᒪᓚᐅᖅᓯᒪᔪᖕᒐ ᓯᕐᖓᒍᒪᓇᖕᒐ ᖃᕐᒪᒦᓐᓂᖓ ᐱᐅᔪᐊᔾᔫᒻᒪᑦ. ᑭᓯᐊᓂ ᒐᕙᒪᒃᑯᑦ ᑕᐃᒫ ᐱᖁᔨᓚᐅᖅᓯᒪᒻᒪᑕ ᐃᑦᓗᑖᕐᓗᑕ ᑭᓯᐊᓂ. ᕿᑐᕐᖓᒐᖃᓕᖅᑎᑦᑐᖕᒐ ᐱᖕᒐᓯᓂ. ᖁᕕᐊᓇᓚᐅᖅᓯᒪᖕᒋᑦᑐᕐᓕ ᐃᑦᓗᑖᕐᓂᐊᓕᖓᑦᑕ, ᖃᐅᔨᒪᖖᒐ ᐃᑦᓗᑦ ᖃᓄᖅ ᐱᓴᐅᔫᖖᒍᒪᖖᒑᑕ.

I began to see many changes when we moved here [to Niaqunnguuq]. We were still living in a *qarmaq* (sod house) when we moved here. The first real change I experienced was moving into a house. I wanted to remain in my qarmaq, but the government forced us to move into a house. I was not happy about this. I had no idea how to live in a house.

ᑭᓯᐊᓂ ᖁᕕᐊᓇᖕᒋᑦᑐᒍᑦ ᑎᑭᒃᑲᓗᐊᖓᑦᑕ ᕿᓄᐃᓵᖓᓯᐊᖃᑦᑕᖁᔭᐅᓯᒪᒐᑦᑕ ᐊᖕᒐᔪᖅᑳᑦᑎᓐᓄᑦ ᑕᒪᒃᑯᐊ ᑭᓯᑐᐃᓐᓇᐃᑦ ᐅᕝᕕᓐᓅᒐᓗᐊᖅᐸᑕ. ᑕᐃᒪᐃᒐᓯᐊᑐᐃᓐᓇᓚᐅᖅᓯᒪᔪᖕᒐᓚ. ᑐᐊᕕᓗᐊᖓᓯᐊᖃᑦᑕᖁᔨᖕᒋᑦᑐᖕᒐ ᐃᓯᒪᓂᑦ ᑎᑭᑕᐅᒐᓗᐊᑭᔅᓯ ᐊᔅᓯᕆᖕᒋᑦᑐᓄᑦ. ᐊᔅᓯᐅᒍᓐᓃᖓᓗᐊᖅᐸᑦ ᐊᑐᖅᑕᓯᑦ, ᐊᑐᖅᑕᐅᔪᓄᑦ ᐊᓯᑦᑎᓐᓄᑦ. ᐊᔅᓯᒥᑕᓇᕚ ᑕᕙ ᐊᑐᖅᑕᐅᖁᔨᐅᔪᖅ, ᐃᒪᐃᑕᐅᖅᑐᖃᖓᓗᐊᖅᑎᑦᓗᒍ ᑭᓯᐊᓂ ᐅᐃᒪᔫᓇᐊᖃᖕᒋᑦᑐᓯ. ᐅᖃᐅᔾᔪᐅᖃᑦᑕᔨᑕᑦᑕᓕ ᐊᖕᒐᔪᖅᑳᑦᑎᓐᓄᑦ ᐱᓵᖅᑐᑦ ᖃᓄᐃᑦᑐᖅᑳᖅᑐᓂ ᐅᖃᖅᐸᔾᔪᒪᑕ. ᑕᐃᒪᐃᑐᐃᓐᓇᓗᑦᑕ ᑲᐸᐲᖅᓯᒪᑦᓗᐊᖃᑦᑕᔪᒪᑦᑕ ᐊᓯᒥᑕᖓᓗᐊᖅᑎᑦᓗᒍ ᐃᓅᓯᖅ. ᐃᑦᑐᑖᖅᑐᑕᓗ, ᐃᑯᒪᖃᓕᖅᑐᑕ ᐊᒻᒪᓗ ᐅᐊᓴᖅᕕᖃᓕᖅᑐᑕ ᐅᖃᓚᐅᑎᑖᖅᐸᑦᓕᐊᑦᑐᑕᓗ. ᑕᒪᒃᑯᐊ ᑭᓯᐊᓂ ᖁᕕᐊᓇᓚᐅᖓᓗᐊᖅᑐᑎᑦ ᖁᕕᐊᓇᕈᓐᓃᖅᐸᓪᓕᐊᓕᓚᐅᖅᓯᒪᒥᔪᖅ ᐊᑭᑦᑐᖅᐸᓪᓕᐊᓂᐊᓗᖕᒋᑦ ᐱᔪᑎᒋᑦᑐᒋᑦ.

I always tried to be prudent with how I encountered the many changes that followed that one. Our parents taught us not to rush into making decisions. They told us to take our time and to have clear minds. This is how I kept my composure through the many changes that occurred. We were warned about the changes to come, and I think this is why I was never shocked by them. Along with the house, soon came washing machines, electricity, and the telephone. Although these changes were fascinating at first, they also came with financial burdens.

ᐅᑦᑐᒥᑦ ᐊᓯᔾᔨᖅᐸᓪᓕᐊᖏᓐᓇᖅᑐᑦ. ᑕᐃᒪᐃᒃᑲᓗᐊᖅᑎᓪᓗᒍ ᒪᑭᑕᒐᓯᐊᖏᓐᓇᖅᓯᒪᔫᒍ. ᐅᑦᑐᒥ ᓵᓚᐅᒐᓯᐊᖅᑲᑦᑕᖏᑦᑐᒍ ᐱᕙᓪᓕᐊᔫᒐᓗᐊᓄᑦ, ᖃᓪᓗᓈᓃᙵᖅᑑᒐᓗᐊᓂ ᐅᕙᒍ ᐃᓅᓯᓇᑦᑕᐅᑎᒥᙵᒥᑕᖕᒥᓐᓂᑦ. ᑭᓯᐊᓂ ᑕᒪᒃᑯᓄᖓ ᐅᐸᓗᐃᔭᐅᒥᐊᖅᑲᖕᒥᑦᑐᓯ, ᓄᑲᖅᑏ ᐅᕙᓐᓂᑦ ᐅᖅᑲᐅᑎᒥᐊᖅᐸᔅᓯ ᑕᐃᒪᐃᑦᑐᓄ ᑎᑭᑕᐅᕇᕐᕋᓗᐊᕈᒃᓯᑦ ᐅᐸᑰᖅᑲᑦᑕᐃᓕᒋᔅᓯ ᐃᓅᓯᑦᓯᓐᓂᑦ, ᑕᐃᒪᐃᖅᑲᑦᑕᖅᑯᔭᐅᓯᒪᒐᑦᑕ.

I am telling the younger people now that they must be prepared for change. Do not get lost in the rapid changes in our lives. Many things come from the Qallunaat that are not in our traditional customs. We must be prepared for these changes. Be prepared as we have been taught to be prepared.

ᐋᓚᓯ ᔫᒥ
ᓂᐊᖁᙳᖅ, ᓄᓇᕗᑦ

Aalasi Joamie
Niaqunnguuq (Apex), Nunavut

2007

ᓇᓗᓇᐃᖅᑯᓯᑦ ᓄᓇᕐᐅᐸ ᐊᑑᑎᖃᕐᓂᖏᑦ ᐊᑎᖏᓐᓄ
Summary of Plant Uses and Names

ᐃᓄᒃᑎᑐᑦ Inuktitut	ᖃᓪᓗᓈᑎᑐᑦ English	ᐊᑑᑎᖏᑦ Uses	#
ᐊᓚᒃᓴᐅᔭᐃᑦ Alaksaujait	Net-Vein Willow	ᑏᒧᐊᖅ, ᓂᕿ, ᐃᑭᐊᖅᑎᒃᓴᖅ, ᒪᓂᖅ Tea, food, insulation, wick	40
ᐃᔨᓯᐅᑎ Ijisiuti	River Algae	ᐊᑲᐅᓯᕵᐅᑎᑦ Medicine	102
ᑲᓪᓚᐃᑦ ᑲᓪᓚᖁᑎᓪᓗ Kallait Kallaqutillu	Bearberry	ᓂᕿᑦ, ᑏᒧᐊᑦ Food, tea	70
ᑭᒍᑕᖏᕐᓇᐃᑦ ᓇᖁᑎᓪᓗ Kigutangirnait Naqutillu	Blueberry	ᓂᕿᑦ Food	73
ᒪᓕᒃᑳᑦ Malikkaat	Mountain Avens	ᓯᓚᒧᑦ, ᓄᓇᐅᑉᓗ ᑭᓯᑦᑎᐊᕐᓂᖓᓂ ᖃᐅᔨᒪᔪᐃᑦ Navigation, weather, medicine	90
ᒪᓂᖅ Maniq	Lamp Moss	ᒪᓂᖅ, ᐊᑲᐅᓯᕵᐅᑎᑦ Wick, medicine	26
ᓂᕐᓇᐃᑦ Nirnait	Snow Lichen	ᓂᕿᑦ, ᐊᑲᐅᓯᕵᐅᑎᑦ Medicine	106

ᐃᓄᒃᑎᑐᑦ Inuktitut	ᖃᓪᓗᓈᑎᑐᑦ English	ᐊᑐᕐᑎᖏᑦ Uses	#
ᐸᐅᓐᓇᐃᑦ Paunnait	Dwarf Fireweed	ᓂᕿᑦ, ᑏᖖᒍᐊᑦ, ᐊᒃᑐᓯᕙᐅᑎᑦ Tea, food, medicine	44
ᐸᐅᕐᖓᐃᑦ ᐸᐅᕐᖓᖁᑎᓪᓗ Paurngait Paurngaqutillu	Crowberry	ᓂᕿᑦ Food	66
ᐳᐊᓗᖖᒍᐊᑦ Pualunnguat	Arctic Cotton	ᒪᓂᖅ, ᓴᓗᒻᒪᖅᓴᐅᑎᑦ, ᐃᑭᐊᖅᑎᒃᓴᑦ Wick, cleaner, insulation	20
ᐳᔪᐊᓗᒃ Pujualuk	Dried Mushroom	ᐊᒃᑐᓯᕙᐅᑎᑦ Medicine	110
ᕿᔪᒃᑖᖅᐸᐃᑦ Qijuktaaqpait	Labrador Tea	ᐊᒃᑐᓯᕙᐅᑎᑦ Medicine	78
ᖁᐊᕋᐃᑦ Quarait	Snow-Bed Willow	ᓂᕿᑦ Food	36
ᖁᖑᓖᑦ Qunguliit	Mountain Sorrel	ᓂᕿᑦ, ᐊᒃᑐᓯᕙᐅᑎᑦ Food, medicine	52
ᓴᐸᖓᕋᓛᖖᒍᐊᑦ ᑐᖅᑕᐃᓪᓗ Sapangaralaannguat Tuqtaillu	Alpine Bistort	ᓂᕿᑦ Food	58

ᐃᓄᒃᑎᑐᑦ **Inuktitut**	ᖃᑦᓗᓈᑎᑐᑦ **English**	ᐊᓪᓚᑎᖏᑦ **Uses**	#
ᓯᐅᕋᐅᑉ ᐅᖃᐅᔭᖏᑦ Siuraup Uqaujangit	Seaside Bluebells	ᓂᕿᑦ, ᐊᑲᐅᓯᓴᐅᑎᑦ Food, medicine	84
ᐅᖅᐱ ᓱᐳᑎᓪᓗ Uqpi Suputillu	Arctic Willow	ᑏᖑᐊᑦ, ᓂᕿᑦ, ᐊᑲᐅᓯᓴᐅᑎᑦ, ᐃᑭᐊᖅᑎᒃᓴᑦ, ᓚᓂᖅ Tea, food, medicine, insulation, wick	30
ᐅᕐᔪ Urju	Peat Moss	ᖁᑦᑕᖅᐱᑦ, ᐃᑭᐊᖅᑎᒃᓴᑦ, ᓚᓂᖅ Diapers, insulation, wick	96

ᓄᓇᕋᐃᑦ
Plants

ᐊᔾᔨᓕᐅᑎᓄᓯᒪᔪᖅ ᐹᑐᓂᒃ ᓕᑐ
Photo by Patrick Little

ᐳᐊᓗᙳᐊᑦ
Pualunnguat / Arctic Cotton

ᐳᐊᓗᙳᐊᑦ ᓄᓇᕗᒻᒥᑦ ᓇᒥᑐᐃᓐᓇᖅ ᓇᓂᔭᐅᔪᓐᓇᖅᑐᖅ. ᐃᓕᓴᕆᔭᕐᓂᖅᑐᐃᑦ ᐱᕈᕇᖅᓯᒪᑎᓪᓗᒋᑦ ᐱᕈᖅᓯᐊᖏᑦ ᖅᑲᒃᑰᒻᒪᑕ ᑕᑭᔪᓂᑦ ᐱᕈᖅᕕᖅᑲᖅᑐᑎᒃ, ᐊᒥᑦᑐᑯᑖᑯᓗᓐᓂᑦ. ᐳᐊᓗᙳᐊᑦ ᐊᔾᔨᒌᖕᒋᑦᑑᑕᐅᒻᒪᑕ ᐃᓚᖏᑦ ᐳᐊᓗᙳᐊᖅᑲᖅᑲᑦᑕᖅᑐᑦ ᐊᑕᐅᓯᕐᒥᑦ ᐃᓚᖏᓪᓗ ᐊᒥᓲᓂᖅᓴᓂᑦ. ᐱᕈᕇᖅᑲᑦᑕᖅᑐᑦ ᐊᐅᔭᖁᑖᒃᑯᑦ ᐅᑭᐊᒃᓴᖁᓯᓐᓂᖕᖓᓂᑦ, ᑎᑦᑕᐅᓇᓱᓐᓂᖏᓐᓂᑦ ᐱᕈᖅᓯᐊᒃᓴᖏᑦ, ᐊᓯᐊᓅᕈᓘᔭᖅᑐᑎᑦ. ᐱᕈᖅᕕᖏᑦ ᑕᑭᓂᖅᑲᕈᓐᓇᖅᑐᑦ 15-30ᓯᐊᓐᑎᒦᑐᓂᑦ. ᑲᑎᒻᒪᒡᒍᔪᐃᑦ,ᐊᒻᒪᓗ ᒪᓴᕈᔪᓪᒦᒡᒍᔪᐃᑦ ᓄᓇᔭᒥᑦ.

Pualunnguat is found across Nunavut. It is very easy to recognize when it is in bloom—its blossoms are bright white puffs that grow on the tall, slender stalks. Some varieties of pualunnguat have one blossom per stalk and others have several. Its blossoms appear early in the summer and remain until fall, when the silky strands of the puffs are blown away on the wind, distributing the seeds. The stalks of pualunnguat grow from about 15 cm to 30 cm high. It tends to grow in large patches, usually across wetter parts of the tundra.

ᐳᐊᓗᙳᐊᑦ ᐊᑐᖅᑕᐅᒐᔪᑦ ᒪᓂᒃᓴᐅᓪᓗᑎᑦ ᖁᓪᓕᕐᒧᑦ. ᒪᓂᓕᐅᕐᓗᓂ ᐃᑯᐊᓚᑦᑎᐊᕐᓂᐊᖅᑐᒥᒃ ᐊᑯᓂᒃᑯᑦ, ᐳᐊᓗᙳᐊᑦ ᒥᖅᑯᖏᓐᓂᑦ ᒪᓂᕐᒧᑦ ᐃᓚᓕᐅᔾᔨᓗᑎᑦ ᓇᓕᒧᕇᓪᓗᑎᑦ (ᒪᓂᖅ; p. 26). ᑲᑎᓪᓗᒋᑦ ᐳᐊᓗᙳᐊᑦ ᒪᓂᕐᓗ ᐊᒃᓴᓕᓪᓗᒋᑦ ᐊᒡᒐᖓᓄᑦ ᑲᑎᑦᑎᐊᕐᓂᐊᕐᒪᑕ. ᐱᑕᖅᑲᖅᑎᓐᓇᒋᑦᓕ ᒪᓂᖅ ᐳᐊᓗᖑᐊᓪᓗ, ᐃᓇᖏᖅᑕᐅᔪᓐᓇᖅᑐᑦ ᓱᐳᑎᓄᑦ (p. 30). ᑕᐃᒪᑦᑕᐅ ᐊᑐᕋᒃᓴᐅᒐᓗᐊᖅ ᐅᕐᔪ (p. 96), ᕿᓯᐊᓂ ᐱᐅᒋᔭᐅᖏᓐᓂᖅᓴᐃᑦ ᐃᑭᑦᑐᑎᒃ ᓄᖑᓴᕋᐃᓗᐊᕐᒪᑕ.

The main use of pualunnguat is as the wick of a *qulliq* (soap stone lamp). To make a wick that will burn evenly and last long, use the silky strands of pualunnguat with an equal amount of *maniq* (lamp moss; p. 26). Roll the pualunnguat and maniq between your palms to combine them. When maniq or pualunnguat is unavailable, either can be substituted with *suputit* (see *uqpi suputillu*, Arctic willow; p. 30). Another possible substitution is *urju* (peat moss; p. 96), but this is not preferred because urju burns too quickly.

ᐳᐊᓗᙳᐊᑦ ᐊᒥᓱᓂᑦ ᐊᓅᑎᓕ ᓴᓗᒻᒪᖅᓴᐃᔭᑎᑦᑎᐊᕆᐅᑦᓗᓂᓗ ᐊᑲᐅᓯᓴᐅᑎᖅᑳᖅᔪᑦᑐᑦᓂᓗ. ᐋᓚᓯᓕ ᐃᖅᑳᐅᒪᒍᖅ ᓄᑕᕋᖕᑎᑕ ᖃᓚᓯᖕᖏᓐᓇᓄᑦ ᓴᓗᒻᒪᖅᓴᐃᔾᔨᑎᕆᖅᑲᑦᑕᓚᐅᖅᓯᒪᕙᒥᐅ. ᓄᑕᕋᖅ ᐃᓅᕋᑖᖅᑎᓪᓗᒍ, ᖃᓚᓯᖓ ᒪᒥᕙᓯᐊᕐᓂᖓᓂᑦ ᒪᖅᑭᑯᔪᓯᕌᓘᒻᒪᑕ. ᐳᐊᓗᐊᙳᐊᖅ ᐅᓐᓄᐊᓕᒦᖅ ᐃᓕᓯᒪᒍᓂ, ᒪᖅᑭᑯᔭᒃᑐᓂᑦ ᓴᓗᐊᖅᑎᑦᑎᒻᒪᑦ

ᐳᐊᓗᙳᐊᑦ ᑕᓯᐅᑉ ᑭᑦᓕᖓᓂ ᓂᐊᖁᙳᒥᑦ.

Pualunnguat at the edge of a pond in Niaqunnguuq (Apex).

ᓴᓗᒻᒪᖅᓴᐃᔾᔪᑎᑦᑎᐊᕚᐅᓪᓗᓂ. ᑕᐃᒫᑦᑕᐅ, ᐳᐊᓗᙳᐊᑦ ᐊᑐᕋᒃᓴᐅᕗᑦ ᓯᐅᓯᓇᔪᒧᑦ ᐃᓐᓇᕐᒧᑦ ᓱᕈᓯᕐᒧᓘᓐᓃᑦ. ᓯᐅᑎ ᒪᕐᑭᒍᓂ,ᐳᐊᓗᙳᐊᑦ ᑕᐳᑦᑖ ᐊᑐᕋᖅᓴᐅᔪᖅ ᐱᕈᕐᕕᖓ ᑕᒪᓐᓇ ᐊᑕᓗᓂ,ᓘᕐᓗᑕ ᓯᐅᑎᑕᓇᔾᔪᑎᒍᑦ. ᐳᐊᓗᙳᐊᑦ ᓴᓗᒻᒪᖅᓴᐃᔾᔪᑕᐅᔪᓐᓇᕐᒥᔪᑦ ᑭᑕᖅᑐᒧᑦ ᒪᑦᑐᑎᑕᐅᓚᐅᖕᒋᑎᓪᓗᒍ ᓯᑕ. ᐳᐊᓗᙳᐊᑦ ᑲᒥᐅᑉ ᐊᑯᓐᓂᖕᒐᓃᑎᑕᐅᖅᑲᑦᑕᖅᑐᕕᓃᑦ ᐊᑐᐃᓐᓇᐅᒐᔭᕐᒪᑕ ᐊᑐᖅᑕᐅᒋᐊᖅᑲᑕᕈᑎ.

Pualunnguat has many uses as a general cleaning cloth with mild medicinal properties. Aalasi recalls using it to clean the umbilical cords of her babies. After a baby is born, there may be some secretion from the cord before it heals. If the pualunnguat is left on overnight, it will absorb any secretion and it will keep the area clean. Similarly, pualunnguat can be used to clean ear infections in adults and children. If there is any leakage from the ear, use a puff of pualunnguat with the stem still attached, like a "Q-tip" or cotton swab. Pualunnguat can also be used to clean wounds before they are mended. People kept puffs of pualunnguat folded into the liners of their kamiks for easy access whenever it was needed.

ᐋᑕᓯ ᐃᖅᑲᐅᒪᔪᖅ ᐊᖕᒐᔪᖅᑳᖕᒋᒃ ᓄᐊᑦᑎᖅᑲᑦᑕᕐᓂᕐᒪᑎᒃ ᐳᐊᓗᙳᐊᓂᑦ ᐊᑭᓯᑕᐊᓇᔭᐅᓂᐊᕐᒥᒃ ᖃᓂᒪᔪᒧᑦ. ᓂᕈᒥᓐᓂᑯᓗᖕᒋᑦ ᐊᑲᐅᓯᔾᔪᑕᐅᔪᓐᓇᕐᒪᑕ ᐊᑲᐅᑦᑎᐊᖕᒋᑦᑐᒧᑦ,ᐱᓗᐊᖅᑐᑎᒃ ᓂᐊᖁᙳᔪᓄ. ᒥᖅᑯᖕᒋᑦᑕᐅ ᐅᖅᑯᔪᑯᓗᓐᓂᑦᐊᓪᓚᓂᓚᐊᖕᒍᔪᓐᓇᕐᒥᔪᑦ ᐃᓅᖅᑖᕐᒧᑦ ᐊᑐᖅᑕᐅᓂᐊᕐᓗᓂ. ᒥᖅᑯᖕᒋᑦ ᖃᑦᑐᓈᖅᑕᐅᑉ ᐊᑯᓂᖕᒐᓄᑦ ᒥᖅᓱᖅᑕᐅᓗᑎᑦ ᓅᕈᓘᔭᕈᓐᓇᐃᑕᒐᔭᖅᑐᑦ. ᑕᐃᒫ ᓴᓇᓗᒍ, ᐳᐊᓗᓄᑦ ᑲᒥᓐᓄᓘᓐᓃᑦ ᐃᓗᑦᑕᑕᐊᖕᒍᔭᓐᓇᖅᑐᑦ.

ᓄᓇᔭᒥᑦ ᑕᑯᔭᕐᓂᖅᑐᑯᓗᐃᑦ ᐳᐊᓗᙳᐊᑦ
ᖃᑯᖅᑕᐃᑦ ᒥᖅᑯᖕᒋᑦ.

The white puffs of *pualunnguat* are easy to spot across the tundra.

Aalasi recalls her parents collecting pualunnguat to make a pillow for someone who was sick. The silky strands have gentle healing properties that help with any ailment, but especially with headaches. The puffs can also be used to make a soothing and well-insulating mattress for an infant. The puffs can be loosely quilted between two layers of cloth to keep it in place. In this same way, it can also be used as a liner for mittens or kamiks.

ᐳᐊᓗᙵᐊᑦ ᓄᐊᓐᓂᐊᕐᓗᒋᑦ ᐱᐅᓛᖑᕗᖅ ᐊᐅᔭᑐᖅᑲᐅᓕᖅᑎᓪᓗᒍ ᒥᖅᑯᒃᑭᑦᑎᐊᕚᓗᑎᓪᓗᒋᑦ. ᑭᓯᐊᓂ, ᐊᑐᓇᐊᖅᑲᕐᒪᑕ ᓯᕗᓂᐊᒍᑦ ᓄᐊᒐᒃᓴᑦᑎᐊᖑᒻᒥᔪᑦ. ᓄᐊᑕᐅᓯᒪᓕᖅᑎᓪᓗᒋᑦᑕ ᐳᐊᓗᙵᐊᑦ ᒥᖅᑯᖏᑦ, ᐅᑭᐅᓕᒫᕐᒧᑦ ᐊᑐᕋᒃᓴᐃᑦ. ᓄᐊᓯᒪᔪᑦ ᐳᐊᓗᙵᐊᑦ ᐆᒪᑯᑖᑦᑎᑕᒃᓴᐅᒻᒥᔪᑦ ᐹᖅᓯᒪᓗᒋᑦ ᐃᒻᒫᕐᔪᒃᓯᒪᓗᒋᑦᓗ.

The best time to collect pualunnguat is in the late summer when it is very fluffy. But, if necessary, it can be collected earlier. Once the puffs of pualunnguat have been collected, they can be kept and used all winter. The puffy blossoms and stalks can also be kept alive for a long time if they are placed in a tin or bucket with a small amount of water at the bottom.

ᐃᖅᑲᓗᐃᑦ ᑰᖓᓂ ᐳᐊᓗᙵᐊᑦ ᐊᒥᓱᐃᑦ (ᓯᐅᑦᕖᐊ ᒍᓇᓂᐅᑦ).

A marsh of *pualunnguat* near Iqaluit Kuunga (the Sylvia Grinnell River).

ᒪᓂᖅ

Maniq / Lamp Moss

ᒪᓂᖅ ᐱᕈᕐᓂᖅᓴᐅᖃᑦᑕᖅᑐᑦ ᓄᓇᕗᒻᒥᑦ ᐅᖅᑰᓂᖅᓴᓂᑦ ᐊᑦᑎᓐᓂᖅᓴᒥᑦ, ᐱᑕᖃᙱᓐᓂᖅᓴᐅᓪᓗᓂ ᖁᑦᑎᓐᓂᖅᓴᒥᑦ. ᒥᑭᑦᑐᑯᓘᓪᓗᑎᑦ ᑲᑎᒪᔪᕚᔾᔭᑰᓪᓗᑎᑦ ᐱᕈᖃᑦᑕᖅᑐᑦ ᖃᐅᓯᐅᓪᓗᑎᓪᓗ.

Maniq is a moss that is common in the more southern parts of Nunavut and less common in the High Arctic. It grows as a moist, spongy carpet made up of very small mounds.

ᐊᑑᑎᓪᓗᐊᙵ ᒪᓂᐅᑉ ᓇᓕᒍᙵᓂᑦ ᑲᑎᑦᑎᓗᓂ ᐳᐊᓗᙳᐊᒍᑦ (p. 20) ᐅᕝᕙᓘᓐᓃᑦ ᓱᐳᑎᒍᑦ (ᐅᖅᐲᑦ ᓱᐳᑎᓪᓗ;p. 30)ᖁᓪᓕᕐᒧᑦ ᒪᓂᐅᓂᐊᕐᓗᑎᑦ.ᐳᐊᓗᙳᐊᑦ ᓱᐳᑎᓪᓗ ᐱᑕᖃᙱᒃᑯᑎᑦ,ᒪᓂᖅ ᑲᑎᑕᒃᓴᐅᕙᑦ ᐅᕐᔪᒧᑦ (ᐅᕐᔪ; p. 96) ᒪᓂᓕᐊᕆᓗᒋᑦ, ᑭᓯᐊᓂ ᐅᕐᔪ ᒪᓂᒃᓴᒧᑦ ᐃᓚᓕᐅᑎᓯᒪᓗᒍ ᓱᒃᑲᓂᖅᓴᒥᑦ ᐃᑭᒃᑲᔭᖅᑐᖅ ᐳᐊᓗᙳᐊᑦ ᓱᐳᑎᐅᓪᓗᓐᓃᑦ ᓴᓂᐊᓂ, ᑕᐃᒪ ᐊᑐᕈᒥᓇᙱᓐᓂᖅᓴᐅᔾᔭᑎᒋᕚ.

The main use of maniq is to combine it in equal parts with either *pualunnguat* (Arctic cotton; p. 20) or *suputit* (*uqpi suputillu*, Arctic willow; p. 30) to make a wick for a *qulliq* (soap stone lamp). If neither pualunnguat nor suputit is available, maniq can also be combined with *urju* (peat moss; p. 96) to make a wick, but a wick made with urju will burn more quickly than a wick made of pualunnguat or suputit, so it is less desirable.

ᒪᓂᖅ ᓲᕈᓗᒃᑐᒧ ᐊᑲᐅᓯᔾᔭᑕᐅᔭᓐᓇᕐᒥᔭᖅ ᐅᖅᓱᐃᔭᐃᒻᒪᑦ. ᓲᕈᓗᒃᑐ, ᓂᕆᔭᓐᓇᖅᑐᖅ (ᑑ ᑖᓚᑎᑐᑦ ᐊᖏᑎᒋᔭᒥᑦ). ᑕᒧᐊᑦᑕᐃᓕᓗᒍ. ᐸᓚᐅᒑᕐᒥᑦ ᓂᕆᖅᑲᓯᐅᔾᔨᖁᔭᖅ ᐄᔭᕐᓂᕐᓂᖅᓴᐅᓂᐊᕐᒪᑦ.

Maniq can also be used to help with heartburn as it absorbs the excess fat in the stomach causing the heartburn. To treat heartburn, swallow a small bunch (about the size of a toonie). Do not chew it. If necessary, combine it with bannock to make it easier to swallow.

ᐋᓚᓯ ᐃᖅᑲᐅᒪᔪᖅ ᒪᓂᖅ ᐊᑐᖅᑕᐅᖃᑦᑕᓚᐅᕐᓂᖏᓐᓂᑦ ᖃᒧᑎᒧᑦ ᐱᐊᒃᓴᐅᑎᓕᐅᖅᑐᓄᑦ ᓂᓚᒻᒥᑦ. ᐃᖅᑲᐅᒪᑦᑎᐊᖅᑐᖅ ᐊᑖᑕᖓ ᖃᒧᑎᒥᓂᑦ ᐱᐊᒃᓴᐃᑎᓪᓗᒍ ᒪᓂᖅ ᖃᐅᓯᖅ ᖃᒧᑎᖏᓐᓄᑦ ᐊᑦᑐᖅᑕᖅᑎᕝᕙᓗᒃᑐᓂᐅᑉ.

Aalasi also recalls that maniq can be used as a sponge to put a new layer of ice on a dog sled runner. She recalls her father working on their sled runners and the quiet rush of the wet maniq passing along them.

ᐋᓚᓯ ᒪᓂᕐᓂᑦ ᐲᖅᓯᔪᖅ ᐃᖃᓗᐃᑦ ᑰᖓᑕ ᓴᓂᐊᓂᑦ (ᓯᐅᓪᕕᐊ ᒍᕆᓂᐅᓪ).

Aalasi scoops up a mound of *maniq* near Iqaluit Kunnga (the Sylvia Grinnell River).

ᐋᓚᓯᐅᑉ ᓱᕋᑦᑎᖅᑕᖓ ᒪᓂᖅ ᐃᓚᓕᐅᑎᓂᐊᕋᒥᐅ ᐳᐊᓗᙱᐊᓄᑦ,ᐃᑯᐊᓚᑦᑎᐊᓲᕐᒥᑦ ᓄᖑᓴᕋᐃᖏᑦᑐᒥᒃ ᒪᓂᔅᓴᓕᐅᖅᑐᖅ.ᒪᓂᖅ ᖁᓪᓕᕐᒧᑦ ᒪᓂᔅᓴᐅᒍᔪᑦ, ᑭᓯᐊᓂᑦᑕᐅ ᐊᑐᖅᓴᐃᑦ ᓲᕿᓗᓗᐊᖅᑐᒧᑦ ᓂᕆᔭᐅᓗᑎᑦ.

Aalasi breaks *maniq* apart to combine it with a handful of *pualunnguat,* preparing an ideal mixture for a slow-burning *qulliq* wick. Maniq is primarily used as a wick for a qulliq, but it can also help with heartburn when swallowed.

ᐅᖅᐱ ᓲᐳᑎᓪᓗ

Uqpi Suputillu / Arctic Willow

ᓲᐳᑎᑦ ᖃᑯᖅᑕᐅᓂᑯᓗᒥᕙᐃᑦ ᓄᓇᕐᐅᑉ ᐱᕈᖃᑦᑕᖅᑐ ᐅᐱᙱᒫᖑᕋᑖᖅᑎᓪᓗᒍ. ᐅᖅᐱᐅᓂᖓᓕ ᕿᔪᑯᑖᑯᓗᒥᕚ. ᑖᓐᓇ ᑎᑎᕋᖅᓯᒪᔫᑉ ᐊᑎᖓ ᐅᖅᐱ ᓲᐳᑦᓘᕗᑦ ᐱᔾᔪᑎᒋᓪᓗᒍ ᐊᑐᓂ ᐅᖃᐅᓯᐅᓂᐊᕐᒪᑏᒃ. ᓄᓇᕐᐅᑉ ᐃᓚᐃᓐᓇᖓ ᑕᐃᔭᐅᒃᑐᔪᖅ ᐅᖃᐅᓯᐅᑎᓪᓗᒋᑦ ᐱᔾᔪᑎᒋᓪᓗᒍ ᓄᓇᕐᐅᑉ ᐃᓚᐃᓐᓇᑐᐊᖓ ᐅᖃᐅᑕᐅᕙᓪᓚᑦ.

Suputit is the fluffy white part of this plant that appears on the catkin early in the summer. *Uqpi* is the woody stem part of the plant. This plant chapter is called *uqpi suputillu* ("uqpi and suputit") because both parts of the plant will be discussed together. The words are often used separately to refer to the plant's specific parts.

ᐅᖅᐱ ᓲᐳᑎᓪᓗ ᓄᓇᕗᓕᒫᐸᓗᒻᒥᑦ ᐱᑕᖅᓲᑦ. ᐱᕈᖅᓯᐊᖏᑦ ᐊᖏᓪᓗᑎᑦ ᐊᐅᐸᕈᔪᑦᑐᑦ ᐊᒻᒪ ᑐᙳᔪᖅᑐᑎᑦ ᐊᐅᔭᓖᓴᒃᑯᑦ ᐱᕈᖅᐸᑦᑐᑎᒃ. ᐱᕈᖅᓯᐊᒃᓴᖃᓕᙶᖓᒥᑦ ᖃᑯᖅᑕᑯᓗᒻᒥᑦ ᒥᖅᑯᖃᓕᖃᑦᑕᖅᑐᑦ (ᓲᐳᑦ). (ᓲᐳᑦ). ᐅᖃᐅᔭᕐᒋᑦ ᕿᑦᑐᖅᑐᑎᑦ ᑐᙳᑦᑐᙵᑦᑐᑎᑦ. ᓵᑦᑐᑯᓗᐃᑦ, ᑮᓪᓚᖏᑦ ᑐᑭᓚᐊᑦᑎᐊᖅᑐᑎᑦ ᐊᒻᒪ ᕿᑎᙵᓄ ᐃᒧᕈᔪᕆᐊᖅᓯᒪᖅᑰᕓᓪᓗᑎᑦ. ᖁᖅᓱᖅᑑᓕᖃᑦᑕᖅᑐᑦ ᐅᑭᐊᒃᓵᒃᑯᑦ. ᕿᔫᓂᖓ (ᐅᖅᐱ) ᑲᔫᔪᑦ. ᑖᓐᓇ ᓄᓇᕐᖅ ᐊᖏᓂᖃᕈᓇᖅᑐ 3-25ᓯᐊᓐᑎᒦᑐᒥᑦ. ᓄᓇᒃᑰᖓᖃᑦᑕᖅᑐᑦ ᐱᕈᕐᓂᖏᑦ ᑭᓯᐊᓂᑦᑕᐅ ᐃᓚᖏᑦ ᖁᒻᒧᑐᐃᓐᓇᖅ ᐱᕈᓲᑦ.

Uqpi suputillu is a common shrub in many parts of

Nunavut. It has large pink and green catkins that bloom early in the summer. When the seeds begin to ripen, they are surrounded by soft white fluff (suputit). The leaves are shiny and bright green. They are thin, smooth-edged, and often curled slightly toward the centre. They turn bright yellow in the fall. The stems (uqpi) are woody and brown. This plant can grow from 3 cm to 25 cm high. It can grow along the ground or vertically, toward the sky.

ᐅᖅᐱ, ᕿᔫᑯᑖᖑᓂᓇᕚ, ᐊᑑᑎᖃᖅᑐᖅ ᐊᔾᔩᙱᑦᑐᓄᑦ. ᐃᑭᑕᐅᔪᓐᓇᖅᑐᑦ (ᑎᐱᑦᑎᐊᓇᑦᑐᑯᓘᑦᓗᑎᒃ). ᐊᑕᐅᓯᑯᑖᒃ ᐃᒐᔾᔭᑕᐅᔪᓐᓇᖅᑐ ᐃᑯᒪᒥᑦ ᐊᐅᓚᑦᑎᔾᔭᑎᒋᔭᐅᓗᓂ. ᑲᑎᖅᓱᖅᓗᒋᑦ ᕿᔪᑯᑖᖑᒋᑦ ᐊᓗᖅᓘᔭᐅᑎᓕᐊᓇᔭᒃᓴᐅᒻᒥᔪᑦ ᐊᒻᒪ ᐊᓪᓕᓂᓕᐊᓇᔭᒃᓴᐅᒋᑦᓗᑎᒃ ᕿᓚᒃᓱᖅᓗᒋᑦ ᐊᑦᑐᓈᖅᓄᑦ ᐊᒥᕕᓂᖅᓄᑦᓘᓐᓃᑦ.

Uqpi, the woody stem part of the plant, can be used for many things. The small branches can be burned for fire (with a lovely scent). A single branch can be used to stir or control a fire as well. The branches can also be woven into entrance mats and sleeping pads by tying them together with thread or pieces of skin.

ᐊᐅᔭᕿᑖᖅ ᐱᕈᖅᑐᖕᒋᑦ ᐱᕈᖅᓯᒪᓕᖅᑎᓪᓗᒋᑦ.

New catkins of *uqpi suputillu* at the beginning of summer.

ᐊᐃᕋᖕᖏᑦ ᑕᒧᐊᔾᐅᔪᓐᓇᖅᑐᑦ ᑭᒍᓯᓇᔪᒧᑦ ᐱᔪᓐᓂᕐᓂᐊᕐᒪᑦ. ᐃᑉᐱᓐᓂᐊᔪᓐᓃᕈᑎᖄᕐᔪᒻᒪᑕ. ᐃᐱᓐᓂᐊᔪᓐᓃᕈᑎ ᐊᑐᕈᒪᓗᒍ, ᖄᖕᖓ ᐊᐃᕋᖕᖓᑕ ᐲᔭᕐᓗᒍ ᑕᒧᐊᕐᑯᔪᖅ ᐃᒃᓵᕐᑭᓂᐊᕐᒪᑦ. ᐊᒻᒪᓗᑦᑕᐅ, ᐅᖃᐅᔭᖕᖏᑦ ᓂᓇᕐᑯᔪᖅ, ᕿᔪᒃᓯᓐᓂᕈᔪᒃᑐᐃᑦ.

The roots can be chewed to relieve a toothache. They contain a mild anesthetic. To access the anesthetic, peel the root and chew on it to release the juices. Also, the leaves are edible. They have a mild, woody flavour.

ᓱᐳᑎᑦ, ᖃᑯᕐᓂᖓ ᒥᖅᑯᕈᔪᑯᓗᒃ, ᐊᐅᔭᒃᑯᑦ ᓄᐊᑕᐅᓰᑦ, ᑎᑦᑕᐅᓚᐅᖕᖏᓐᓂᖕᖏᓐᓂᑦ. ᓱᐳᑎᑦ ᒪᓂᓕᐊᓇᔭᐅᒡᒍᔪᐃᑦ ᖁᓪᓕᕐᒧᑦ. ᒪᓂᓕᐊᖕᒍᓂᐊᕐᓗᓂᓕ ᖁᓪᓕᕐᒧᑦ ᐊᑐᖅᑐᒃᓴᐅᓗᓂ, ᓱᐳᑎᑦ ᒪᓂᕐᒧᑦ ᑲᑎᑕᐅᒡᒍᔪᑦ ᓇᓕᒧᖏᓪᓗᐊᑐᐃᓐᓇᕐᓗᒋᑦ. ᐸᐊᓗᖖᒍᐊᑦ (p. 20) ᒪᓂᕐᒧᑦ ᐃᓚᓕᐅᑎᔾᔭᒃᓴᐅᒻᒥᔪᑦ ᒪᓂᓕᐊᖕᒍᓂᐊᕐᓗᓂ. ᑭᓯᐊᓂ ᓱᐳᑎᑦ ᐱᐅᕆᔭᐅᓂᖅᓴᐅᓰᑦ ᓄᖕᒍᓴᕋᐃᖕᖏᓐᓂᖅᓴᐅᒻᒪᑕ. ᒪᓂᖅᑕᖃᖅᑎᓐᓇᒍᓕ, ᓱᐳᑎᑦ ᐃᓚᓕᐅᔾᔭᐅᒃᓴᐅᕗᖅ ᐸᐊᓗᖖᒍᐊᒧᑦ. ᐊᒻᒪᓯᓕ ᑕᐃᒪ ᐅᕐᔪ (p. 26) ᐊᑐᖕᓖᕋᒃᓴᑦᑎᐊᖅ (p. 96), ᑭᓯᐊᓂ ᐅᕐᔪ ᐊᑐᖅᑕᐅᖃᑦᑕᖅᑐᖅ ᐊᓱᒃᓴᖄᖕᖏᑦᑎᐊᖅᑎᓪᓗᒍ ᐱᔾᔪᑎᒋᓪᓗᒍ ᓄᖕᒍᓴᕋᐃᓐᓂᖓ.

ᒥᖅᑯᓕᑯᓗᐃᑦ ᐊᐅᔭᓪᓗᐊᒃᑯᑦ, ᐃᖃᓗᐃᑦ ᐃᒪᖕᖓᓄᑦ ᓵᖖᒐᓪᓗᐊᖅᑐᑦ (ᐃᖃᓗᐃᑦ, ᓄᓇᕗᑦ).

Fuzzy *suputit* in mid-summer, overlooking Frobisher Bay (Iqaluit, Nunavut).

Suputit, the white fluffy part, is collected in the early summer, before it blows away. The main use of suputit is as the wick of a *qulliq* (soap stone lamp). To make the wick for a qulliq, suputit is usually mixed with maniq in equal parts. *Pualunnguat* (Arctic cotton; p. 20) can also be combined with *maniq* (lamp moss; p. 26) to make a wick. However, some prefer suputit because it burns slightly more slowly. When maniq is unavailable, suputit can be combined with pualunnguat. Another possible substitution is *urju* (peat moss; p. 96), but urju is only used when nothing else is available because it burns too quickly.

ᐅᑭᐊᒃᖃᒃᑯᑦ, ᐅᖅᐱᐅᑉ ᓱᐳᑎᓪᓗ ᐅᖃᐅᓯᖏᑦ ᖁᖅᓲᖅᑐᑦ ᑕᖅᓴᐅᑦᑎᐊᖅᑐᐊᓗᐃᑦ ᑳᓪᓚᕿᑎᐅᑉ ᐊᐅᐸᖅᑐᐃᑦ ᐅᖃᐅᓯᖏᓐᓂᑦ (ᐅᖃᐅᓯᖏᑦ ᑳᓪᓚᕿᑎᐅᑉ).

In autumn, the yellow leaves of *uqpi suputillu* stand out amongst red *kallaquti* (the leaves of bearberry).

ᐅᖅᐱ ᓱᐳᑎᓪᓗ ᐊᐅᔭᐅᑉ ᖅᑭᑎᖕᖓᓂᑦ.

Uqpi suputillu in mid-summer.

ᖁᐊᕋᐃᑦ
Quarait / Snow-Bed Willow

ᖁᐊᕋᐃᑦ ᒥᑭᑦᑐᑯᓘᕗᖅ ᐅᑉᐱᒐᐅᓪᓗᓂᑦ ᐊᖏᓂᖓ 0.5ᒥᑦ 5ᓯᐊᓐᑎᒦᑑᔪᓐᓇᖅᑐᖅ. ᐊᒪᓗ ᐸᓗᒃᑐᑦ ᐅᖅᑲᐅᔭᖏᑦ 6ᒥᑦ 21 ᒥᓕᒦᑑᔪᓐᓇᖅᑐᑎᑦ ᑕᑭᓂᖏᑦ. ᑐᖑᑦᑐᓇᑦᑑᔪᑦ ᐅᖅᑲᐅᔭᖏᑕ ᑭᑦᓚᖏᑦ ᖃᑦᑎᒐᓛᑦᑐᑎᒃ. ᐱᕈᖅᓯᐊᖏᑦ ᐊᐅᐸᓪᓚᓇᑦᑐᑯᓗᐃᑦ ᐅᖅᑲᐅᔭᖏᓐᓃᙶᖅᐸᑦᑐᑎᒃ, 5ᒥᑦ 11 ᒥᓕᒦᑐᒥᑦ ᑕᑭᓂᖃᕈᓐᓇᖅᑐᑎᒃ. ᖁᐊᕋᐃᑦ ᐱᕈᒐᔪᑦᑐᑦ ᐊᐳᑎᓂᑦ ᐊᐅᓚᔮᖅᑐᕕᓂᕐᓂᑦ.

Quarait are a very tiny willow that grows from 0.5 cm to 5 cm high. The nearly round leaves grow from 6 mm to 21 mm long. They are dark green with bumpy edges. The flowers are bright red catkins that grow up from the leaves, from 5 mm to 11 mm high. Quarait grow most commonly under late snow beds.

ᖃᐅᕋᐃᑦ ᓂᕿᑦᑎᐊᕙᓪᓚᓇᐅᔪᖅ ᒪᒪᖅᑐᑎᓪᓗ. ᐋᓚᓯ ᐃᖅᑲᐅᒪᔪᖅ ᐅᖅᑲᐅᔭᖏᓐᓂ ᓂᕇᓐᓇᖅᑕᓚᐅᕐᓂᖏᓐᓂᑦ ᓂᕕᐊᖅᓯᐊᖑᑦᓗᓂ. ᐊᒫᖏᑦ ᐅᖅᑲᐅᔭᖏᓪᓗ ᓂᕿᑦᑎᐊᕚᔪᑦ, ᑭᓯᐊᓂ ᑭᑦᓚᐸᖏᑦ ᐊᐅᐸᖅᑐᐃᑦ ᓂᕿᐅᖏᑦᑐᑦ. ᐅᖅᑲᐅᔭᖏᑦ ᑕᐃᓛᑐᐃᓐᓇᖅ ᓂᕆᔭᒃᓴᐃᑦ ᐅᕙᓗᓐᓃᑦ ᐃᓚᓕᐅᑎᓗᒋᑦ ᑲᑎᑕᐹᓄᑦ ᑐᖑᔪᖅᑕᓄᑦ ᐱᕈᖅᑐᕕᓂᕐᓄᑦ. ᐊᓗᓕᐊᓇᔭᒃᓴᐅᒻᒥᔪᑦ. ᐊᑯᑦᑎᓗᒋᑦᓕ ᐊᒡᒐᖕᓄᑦ ᐊᒃᓴᓕᑦᓗᒋᑦ ᐃᔅᓵᕿᓇᓱᓐᓂᖏᓐᓂᑦ; ᐅᖅᓱᒧᑦ ᐃᓚᓕᐅᑎᓕᕐᒥᓗᒋᑦ.

Quarait are delicious and very nutritious. Aalasi recalls eating the leaves constantly when she was growing up. The rhizomes (underground stems) and leaves are both good to eat, but the catkins are not. The leaves can be eaten by themselves or with other greens. They can also be made into an *alu* (pudding). When using the leaves to make an alu, rub them between your palms first to release the juices; then, add them to the fat.

ᖁᐊᕋᐃᑦ ᑏᖖᑐᐊᓕᐊᓇᔭᐅᖃᑦᑕᖕᒋᑦᑐᑦ. ᑭᓯᐊᓂᓕ ᐃᓚᓕᐅᑎᔭᒃᓴᐅᕗᑦ ᐃᖃᓗᖠᓂᓕᐅᖅᑕᓄᑦ.

Quarait are not used for tea. However, they can be added as a seasoning to the cooking water of boiled char.

ᖁᐊᕋᐃᑦ ᒥᑭᓛᖑᕗᑦ ᓇᐹᖅᑐᓕᒫᓂᑦ ᓄᓇᕐᔪᐊᒥᑦ. ᐱᕈᒐᔪᓐᓂᖕᒋᑦ ᐊᐳᑎᖅᑯᑖᓲᓂᑦ ᖃᐅᔨᒪᔭᐅᔪᑦ, ᖃᑦᑐᓈᑎᑐᑦ ᐊᑎᖃᖅᑐᑦ “ᐊᐳᒻᒥᑦ ᐃᓂᖥᑦ ᐱᕈᖅᓯᐊᑦ.”

Quarait are some of the tiniest trees in the world. They are known to grow where the snow remains longer than it does in surrounding areas, so this plant's English name is “snow-bed willow.”

ᖁᐊᕋᐃᑦ ᒥᑭᑦᑐᕋᓛᑯᓗᐃᑦ ᐊᐅᐸᖅᑐᑦ ᐱᕈᖅᑲᑦᑕᖅᑐᑦ ᐅᖅᑲᐅᔭᓂᑦ.

Quarait has tiny red catkins that grow up from the leaves.

ᖁᐊᕋᐃᑦ ᐊᓗᒃ
Quarait Alu

- 1 ᐊᒻᒪ ᑕᑕᓪᓗᒍ ᖁᐊᕋᐃᑦ ᐅᖃᐅᓯᖏᓐᓂ
- 2 ᐊᓗᑎᑯᔅᓯᑦ ᓇᑦᑎᐅᑉ ᐊᐅᖓ
- 1 ᐃᖅᖑᓯᖅ ᓇᑦᑎᐅᑉ ᐅᖅᓱᖓ, ᑭᓱᑐᐃᓐᓇᐅᓪᓘᓐᓃᑦ ᐅᖅᓱᙶᖓᓂᑦ

- 1 handful of quarait leaves
- 2 tablespoons of seal blood
- 1 cup of seal fat or other fat

ᓯᕗᓪᓕᖅ, ᐅᖃᐅᓯᖏᑦ ᐊᒻᒪᓐᓂᑦ ᐊᒃᓴᓪᓗᒋᑦ ᐃᒃᓵᖅᑭᓕᕋᓱᓐᓂᖏᓐᓂᑦ. ᐃᖑᓚᖅᕕᒻᒧᑦ ᑲᑎᓪᓗᒋᑦ ᐅᖃᐅᓯᑦ ᓇᑦᑎᐅᓪᓗ ᐊᐅᖓ. ᐃᓚᓕᐅᑎᕙᓪᓕᐊᕋᔾᔪᓪᓗᒍ ᓇᑦᑎᐅᑉ ᐅᖅᓱᖓ, ᐊᓘᑎᕋᓛᕐᒥᑦ ᐊᑐᕐᓗᑎᑦ, ᐊᑯᑦᑎᐊᕐᓗᒍ. ᓂᕆᑐᐃᓐᓇᕋᒃᓴᐅᔪᑦ ᐃᓛᒃᑯᑦ ᐅᕝᕙᓘᓐᓃᑦ ᐆᔪᖅᑐᓕᕈᕕᑦ ᓂᕆᔭᒃᓴᐅᔪᓐᓇᖅᑐᖅ.

First, prepare the leaves by rubbing them between your palms until the juices are released from the leaves. Combine the leaves with the seal blood in a bowl. Slowly add the seal fat, spoonful by spoonful, combining well. Enjoy alone or as a side dish with meat.

ᐋᓚᓯ ᒪᒪᑕᕈᖅ ᖁᐊᕋᕐᓂᑦ ᔪᓚᐃ ᕿᑎᖓᓂᑦ.

Aalasi enjoying *quarait* in mid-July.

ᐊᓚᒃᓴᐅᔭᐃᑦ

Alaksaujait / Net-Vein Willow

ᐊᓚᒃᓴᐅᔭᐃᑦ ᑐᖑᔭᖅᑕᕈᔭᑯᓗᒻᒥᑦ ᐊᒥᐊᖅᑲᖅᑐᑦ ᐊᒻᒪᓗᑭᓵᕈᔮᓪᓗᑎᒃ. ᐅᖃᐅᓯᖏᑦ ᕿᓪᓕᕈᔭᒃᑐᑎᒃ ᑕᖅᑲᖑᐊᓕᐊᓘᓪᓗᑎᒃ ᓄᓗᐊᑎᑐᑦ ᐋᖅᑭᒃᓱᖅᓯᒪᔭᓂᒃ. ᐱᕈᖅᑲᑦᑕᖅᑐᑦ ᓄᓇᐅᑉ ᖄᖓᒍᑦ. ᐊᒥᐊᖏᑦ ᖁᖅᓱᖅᑑᖅᑲᑦᑕᖅᑐᑦ ᑲᔪᕈᔮᓪᓗᑎᑦ ᐅᕝᕙᓘᓐᓃᑦ ᐊᐅᐸᖅᑑᓪᓗᑎᑦ ᑲᔪᕈᔫᓂᖅᑲᖅᑐᑎᒃ. ᐊᓚᒃᓴᐅᔭᐃᑦ ᑭᓪᓚᐸᑖᑦ ᐊᐅᐸᓪᓚᓇᑦᑐᓂᑦ ᐅᖃᐅᓯᖏᓐᓃᖕᖘᖅᑲᑦᑕᖅᑐᑎᒃ, 4 ᓯᐊᓐᑎᒦᑐᒥᑦ ᐊᖏᓂᖅᓴᐅᓕᕈᓐᓇᕋᑎᑦ.

Alaksaujait are easily recognizable by their shiny, bright green leaves, marked deeply with veins in a net-like pattern. The leaves are almost round, reaching a gentle point at the tip. The branches grow along the ground. They are smooth and can be yellow-brown or red-brown. Alaksaujait have small, dark red catkins that grow up slightly from the leaves (usually no more than 4 cm).

ᑏᖑᐊᑐᐃᓐᓇᐅᓗᐊᖅᑲᑦᑕᖅᑐᑦ ᐊᓚᒃᓴᐅᔭᐃᑦ. ᐅᖃᐅᓯᖏᑦ, ᓇᑲᖏᑦ ᐊᐃᕋᖏᓪᓗ ᓄᐊᑕᐅᒍᓐᓇᖅᑐᑦ ᐊᐅᔭᐅᑉ ᕿᑎᖓᓂᑦ ᓄᖑᐊᓄᑦ. ᐃᓕᒃᑯᐊᓇᔭᐅᖅᑲᑦᑕᕈᓐᓇᖅᒥᔪᑦ ᐅᑭᐅᓕᒫᖅᒍᑦ. ᐋᓚᓯ ᐃᖅᑲᐅᒪᔭᖅ ᐊᖑᓇᓯᑦᑏᑦ ᑕᖅᑯᐊᖅᑲᑦᑕᓚᐅᖅᓂᖏᓐᓂᑦ ᐊᓚᒃᓴᐅᔭᓂᑦ ᐊᐅᓪᓚᖅᓯᒪᓂᖏᓐᓂᑦ ᐱᔭᐅᖅᑲᑦᑕᖅᓂᐊᖅᑐᓂᒃ. ᐅᖃᐅᓯᖏᑦ ᐆᖅᑕᐅᓯᒪᖅᑲᑦᑕᖅᑐᕕᓃᑦ ᕿᓯᔭᓂᑦ ᐆᑲᒻᒪᑐᖕᓄᑦ.

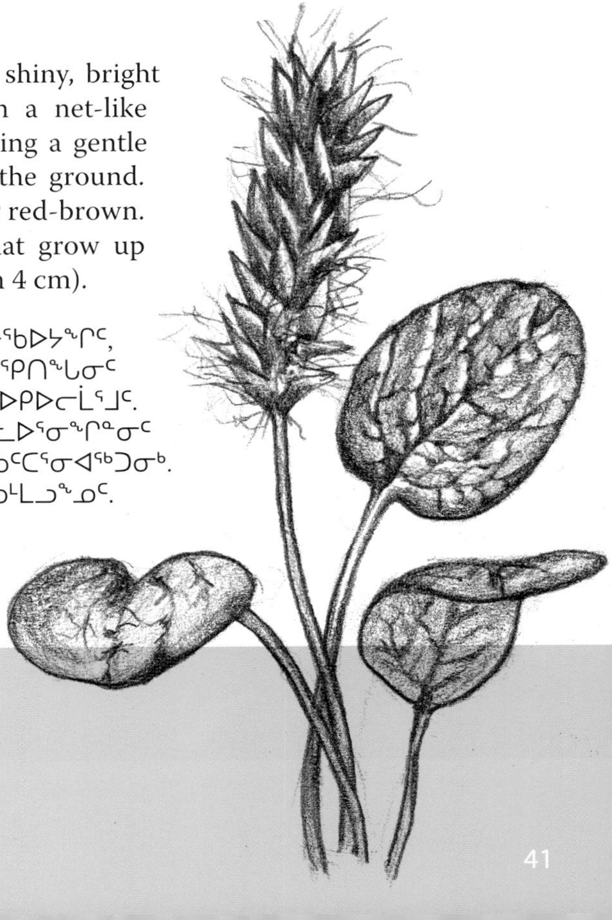

The primary use of alaksaujait is to make tea. The leaves, stems, and roots can be gathered and used from mid to late summer. They can also be stored for later use during the winter. Aalasi recalls that hunters would take alaksaujait with them on their trips to drink while they were away. They would store the leaves in seal skin purses.

ᐊᓚᒃᓴᐅᔭᐃᑦ ᑏᓐᖑᐊᖑᓪᓗᑎᑦ ᓴᙱᔪᒻᒪᓃᑦ ᑎᒥᒧᑦ ᐊᑑᑎᖃᑦᑎᐊᖅᑐᑎᒃ ᐊᒻᒪ ᐊᑲᐅᓯᓴᐅᑎᐅᑦᑎᐊᓇᓪᓗᑎᒃ ᖃᓄᑐᐃᓐᓇᖅ ᐋᓐᓂᐊᔪᓄᑦ ᓈᕐᓗᓐᓇᖅᑐᓄᑦ ᐱᓗᐊᖅᑐᓂ. ᑎᒥᒥᑦ ᐊᐅᒪᔾᔭᑦᑎᒃᑎᓲᖑᔪᑦ, ᑎᒥᒧᑦ ᑕᒪᓐᓇ ᐊᑲᐅᓯᓴᐅᑎᐅᓪᓗᓂ. ᓴᙱᓗᐊᒧᑦ ᐋᒪᒪᑦᑎᑦᑎᔪᖅ ᑏᑐᕐᓂᕈᓂ ᐋᒪᒪᑦᑎᑕᖓ ᓄᑕᕋᖅ ᐊᑦᑐᖅᑕᐅᒐᔭᖅᑐᖅ.

Alaksaujait tea is a strong tea that can help with almost all illnesses, especially illnesses that involve an upset stomach. The tea creates body heat and causes a person to sweat, which is good for their health. This tea is powerful enough that its effects may be felt by a breastfeeding baby whose mother has drunk the tea.

ᑏᓐᖑᐊᓕᐅᕐᓂᐊᕐᓗᓂᓕ ᐊᓚᒃᓴᐅᔭᓂᑦ, ᓄᐊᑦᑎᓚᐅᕐᓗᑎᑦ ᐊᒻᒪᐃᑦ ᑕᑕᓗᐊᕐᓗᒍ ᐅᖃᐅᔭᓂᑦ,ᓇᑲᖏᓐᓂᑦ ᐊᐃᕿᖏᓐᓂᓪᓗ (ᑭᓪᓚᐸᖏᑦ ᐱᖃᓯᐅᑎᖏᓪᓗᒋᑦ). ᐃᒪᖅ ᑎᖅᑎᑎᓪᓗᒍ ᐃᓱᖅᓯᕈᔪᒐᓱᓐᓂᖓᓂ-ᐅᔾᔨᓇᑲᔭᖅᑕᒃᓴᓇᕈᑦ ᑖᓐᓇ ᑏᓐᖑᐊᖑᓗᒍ ᐊᔾᔨᐅᑦᑎᐊᖏᑦᑐᑯᓘᒻᒪᑦ.

ᐊᓚᒃᓴᐅᔭᐃᑦ ᐅᓐᓄᓕᖅᑎᓪᓗᒍ ᐃᖃᓗᐃᑦ ᓴᓂᐊᓂᑦ ᐊᐅᔭᐅᑉ ᕿᑎᙵᓂᑦ.

Alaksaujait at dusk in mid-summer near Iqaluit.

To make tea from alaksaujait, collect a generous handful of leaves, stems, and roots (but not catkins). Boil them until the water has thickened slightly—you will probably notice the unique consistency of tea made from this plant.

ᐊᐃᕋᖕᒋᑦᑕᐅᑉ ᑖᒃᓯᒪᑦ ᐱᕈᖅᑰᑉ ᓂᕿᑦᑎᐊᕙᐅᔪᑦ. ᒪᒪᖅᑐᑯᓘᓪᓗᑎᑦ ᑕᐃᒫᑐᐃᓐᓇ ᓂᕆᒋᐊᖕᒐ.

The leaves and roots of this plant are also nutritious. They are enjoyable to eat fresh and raw.

ᐊᓚᒃᓴᐅᔭᖅ ᑏᑐᕐᓗᒍ
ᑎᒥᒥᑦ ᐊᐅᒪᔾᔭᑦᑎᒃᑎᓰᙶᔪᑦ,
ᐋᓂᐊᒥᑦ ᑕᐃᒪ ᐊᓂᐊᑦᑎᔪᖅ, ᐱᓗᐊᖅᑐᒥᒃ ᓈᕐᓘᓪᓗᓂ.

Tea from *alaksaujait* creates body heat and causes a person to sweat out illnesses, especially those related to the stomach.

ᐊᔾᔨᓕᐅᑎᓇᔭᕕᓂᖅ ᐃᐊᓚᓐ ᓯᒡᓗᕐ
Photo by Ellen Ziegler

ᐸᐅᓐᓇᑦ
Paunnait / Dwarf Fireweed

ᐸᐅᓐᓇᐃᑦ ᐅᑭᐅᖅᑕᖅᑐᒥᑦ ᓇᓂᔭᐅᓲᖑᕗᑦ. ᐱᕈᒪᔪᒃᑐᑦ ᓱᐅᕋᕐᒥᑦ ᓱᕋᒃᑕᐅᓯᒪᔪᓂᑦ, ᐅᔾᔨᕆᒍᓯᖅᑖᕆᓇᔭᑎᑦ ᓄᓇᓕᓐᓂᑦ ᐊᖅᑯᑎᐅᑉ ᑭᒡᓕᖓᓂ. ᓄᓇᓖᑦᓕ ᓯᓚᑖᓂᑦ ᖃᖅᑳᔮᓃᓯᑦ, ᓱᐅᕋᔮᒥᑦ ᔫᓐᓃᑦ ᓯᔾᔭᒥ.ᐱᕈᓕᕖᖅᑎᑦᓱᒍ ᐅᐱᕐᖓᒃᑯᑦ ᐸᐅᓐᓇᑦ ᐊᐅᐸᓪᓚᓐᑐᑯᓘᖅᑲᑕᖅᑐᖅ. ᑕᒻᒪᓕ ᑐᖑᔪᖅᑕᕈᕌᖓᑕ, ᐱᕈᖅᓯᐊᖏᑦ ᐱᕈᓐᖅᑲᑕᖅᑐᑦ. ᑖᒃᑯᐊᓕ ᓇᒥᑐᐃᓐᓇ ᔪᓚᐃ ᕿᑎᖅᐸᓯᖓᓂ ᐱᕈᓐᑎᐊᖅᑲᑕᖅᑐᑦ ᐋᒡᒌᓯᐅᓕᕋᖓ ᑭᓯᐊᓂᑦ ᑐᖁᓕᕝᑐᑎᒃ.

Paunnait are found across the Arctic. It tends to grow where the soil has been disturbed, so you might notice it near roads and waterways around your community. Out on the land, this plant grows on exposed hillsides and on gravelly shores. When it first begins to grow in the spring, the stalks and leaves of paunnait are dark red. Then they turn green and its flowers begin to open. In most regions, the flowers open in mid-July and remain until late August.

ᐸᐅᓐᓇᑦ 5 ᓯᐊᓐᑎᒦᑐᓂᑦ 30ᓯᐊᓐᑎᒦᑐᓄᑦ ᐊᖏᓂᑖᕈᓐᓇᖅᑐᑦ. ᐸᐅᓐᓇᐃᑦ ᐅᖃᐅᔭᖏᑦ ᑐᖑᔪᖅᑕᐅᑦᑐᑎᒃ ᑭᓪᓕᖏᑦ ᓂᕈᒥᑦᑑᑦᑐᑎᒃ. ᐃᓛᓐᓂᒃᑯᑦ ᑕᑭᔪᑯᑖᑯᓘᖅᑲᑕᖅᑐᑦ ᐃᓛᓐᓂᒃᑯᑦᑐ ᓇᐃᑦᑑᑦᑐᑎᒃ ᓯᓕᑦᑐᑲᑦᓚᐅᑦᑐᑎᑦᑐ. ᐱᕈᖅᓯᐊᖑᓂᖏᑦ ᐊᖏᔫᑦᑐᑎᑦ (2-5 ᓯᐊᓐᑎᒻᒦᑐᒦᑐᑎᒃ ᑕᑭᓂᖏᑦ)ᐊᐅᐸᕈᔫᒃᑑᖅᑲᑕᖅᑐᑦ,

ᐊᐅᐸᓪᓚᓇᑦᑐᑎᓪᓗ. ᑎᓴᒪᐅᓕᖓᔪᓂᑦ ᐱᕈᖅᑐᖅᑲᖅᐸᑦᑐᑎᒃ ᐊᒻᒪᓗ ᐊᑦᑕᑎᖏᑦ ᑕᖅᓴᐅᓂᖅᓴᐅᓪᓗᑎᑦ ᐊᐅᐸᕈᔪᒃᑑᓪᓗᑎᑦ. ᐸᐅᓐᓇᐃᑦ ᐊᖏᔫᒻᒪᑕ ᑕᑯᔭᕐᓂᖅᑐᑯᓗᐃᑦ.

Paunnait grow between 5 cm to 30 cm tall. The leaves are bluish-green and have smooth edges. They can be long and narrow and they can also be shorter and oval-shaped. The flowers are large (2 cm to 5 cm across) and they are bright pink or purple. The flowers have four wide petals and four darker pink sepals (sepals are special leaves that enclose the flower bud). Because the flowers are so large and bright, paunnait are easy to spot.

ᐸᐅᓐᓇᐃᑦ ᐊᑑᑎᓕᒻᒪᓈᑦ. ᓂᕐᑭᑦᑎᐊᕙᐅᓪᓗᑎᒃ ᑎᒥᒧᑦ ᐱᐅᓪᓗᑎᒃ.

Paunnait have many uses. They are an all-purpose source of nutrition and are good for overall health.

ᑐᙵᔭᖅᑕᐅᓂᖏᑦ ᓂᕆᔭᐅᔭᓐᓇᖅᑐᑦ ᑕᐃᒫᑐᐃᓐᓇᖅ ᐊᓘᓗᒍᓘᓐᓃᑦ ᐅᖅᓱᒥᒃ ᐊᒻᒪᓗ ᐊᐅᒻᒥᑦ ᐃᓚᓯᒪᓗᒍ.

The green parts of the plant can be eaten plain and they can also be mixed with fats and blood to make nutritious side dishes.

ᐸᐅᓐᓇᐃᑦ ᐱᕈᖅᓯᐊᒃᓴᖅᑲᓕᓰᑦ ᔪᓚᐃ ᐱᒋᐊᕐᓂᖓᓂ.

The buds of *paunnait* in early July.

ᐸᐅᓐᓇᐃᑦ ᐊᒫᖕᑎᑦ ᑕᐃᒫᑐᐃᓐᓇᖅ ᓂᕆᔭᒃᓴᐃᑦ. ᓄᑕᕋᕐᓅᖅᑕᐅᔪᓐᓇᕐᒥᔪᑦ ᐊᒫᒪᕋᐅᔭᕆᔭᐅᑎᓪᓗᒋᑦ. ᓄᑕᕋᕐᒨᖅᑕᐅᓂᐊᕈᓂ ᐊᒫᖅ ᑕᑭᔫᒋᐊᕐᓗᒍ ᑐᐱᔅᔪᑕᐅᔪᓐᓇᖅᑎᒋᐊᖕᒋᓪᓗᒍ. ᓇᐃᑦᓕᓗᐊᓕᕈᓂᓕ ᑭᒻᒪᓕᓂᐅᓗᐊᓕᕈᓘᓐᓃᑦ ᓄᑕᕋᖅ ᐋᖅᑯᔭᖅ. ᐊᒫᖅ ᑭᒍᓯᓴᔪᒥᒃ ᐃᑲᔪᕈᓐᓇᖅᑐᖅ ᐊᒻᒪᓗ ᓂᖅᑭᑦᑎᐊᕙᐅᑦᓗᓂ.

The rhizomes (underground stems) of paunnait can be eaten just as they are. They can also be given to babies as pacifiers. If a rhizome is given to a baby, it should be fairly long so that it is not easy to swallow. Also, it should be taken away before it has been chewed down too much. The rhizomes help with teething and they also provide nutrients.

ᐸᐅᓐᓇᐃᑦ ᑏᖖᒍᐊᑦᑎᐊᕚᓘᕗᑦ, ᑎᒻᒥᑦᑎᐊᕆᓐᓇᖅᑐᑎᒃ. ᑎᒥᒧᑦ ᐊᑑᑎᖅᑲᑦᑎᐊᕐᓂᖕᒐᓄᑦ ᐸᐅᓐᓇ ᑏᖕᒍᓗᒍᑦ ᓄᑕᕋᕐᒧᑦ ᐊᑐᖅᑕᐅᔪᓐᓇᖅᑐᖅ ᐃᒻᒧᒋᔭᐅᖕᒑᕐᓗᓂ ᐃᒻᒧᖅᑲᖕᒋᑦᑐᒧ. ᐃᓄᐃᑦ ᐃᓚᖕᒋᑦ ᐅᑦᓗᑕᒫᖅ ᑏᑐᖅᑲᑦᑕᖅᑐᑦ ᐸᐅᓐᓇᕐᓂᑦ ᑎᒥᑦᑎᐊᕆᒍᒪᑦᑐᑎᑦ. ᐸᐅᓐᓇᐃᑦ ᑏᖖᒍᐊᓕᐅᕐᓂᐊᕈᕕᑦ ᓄᐊᑦᑎᖅᑳᕆᐊᖅᑲᖅᐳᑎ. ᐅᐊᓴᖅᓗᕆᑦ ᓴᓃᔭᑦᑎᐊᖅᓗᕆᑦ, ᐱᕈᖅᓯᐊᖕᒋᑦ ᐲᔭᖅᓗᕆᑦ ᐊᒫᖕᒋᑦᓗ. ᐸᐅᓐᓇᐃᑦ ᐅᖅᑲᐅᔭᐅᓂᖕᒋᑦ ᑎᖅᑎᑎᑦᓗᒋᑦ ᐃᓱᖅᓂᖕᒐ ᓈᒻᒪᒋᓕᖅᓱᓐᓂᖅᓂᑦ. ᒪᒪᑕᓐᓂᐊᖅᐳᑎᑦ!

Paunnait can be boiled into a delicious tea that is very good for overall health. A very strong paunnait tea can even replace milk or

ᐸᐅᓐᓇᐃᑦ ᓇᓗᓇᖕᒋᑦᑎᐊᖅᑐᑦ ᐊᖕᒋᔫᒻᒪᑕ ᐊᒥᐊᖕᒋᓪᓗ ᑕᖅᓴᐅᓪᓗᑎᑦ.

Paunnait are easily recognizable by their large size and bright colour.

formula for babies because it is so high in nutrients. Some people drink this tea every day for good health. To make delicious paunnait tea, gather several of the plants. Rinse them to remove any dust, remove any flowers or rhizomes, and then boil all the leaves until the desired strength of tea is reached. Enjoy!

ᐊᐅᔭᓚᒫᖅ ᐸᐅᓐᓇᐃᑦ ᓄᐊᒐᒃᓴᐅᕗᑦ ᑭᓯᐊᓂᑦ ᐱᐅᓂᖅᐹᖑᕗᑦ ᓄᐊᒋᐊᖓ ᐅᐱᕐᖔ ᓄᖖᒍᑉᐸᓗᐊᒍᑦ ᐱᕈᖅᓯᐊᖏᑦ ᑲᑉᓯᒪᓂᖅᑎᓪᓗᒋᑦ ᐅᖅᑲᐅᔭᖏᑦ ᑐᖑᔪᖅᑕᐅᑎᓪᓗᒋᑦ ᓯᓚ. ᑕᕝᕙ ᑕᐃᒪᐃᑎᓪᓗᒋᑦ ᐸᐅᓐᓇᐃᑦ ᐊᑲᐅᓯᓴᐅᑎᖃᕐᓂᖅᐹᖑᕗᑦ ᐊᒻᒪᓗ ᒪᒪᖅᓯᑦᑎᐊᖅᓯᒪᓪᓗᑎᑦ. ᓄᐊᑕᐅᓚᐅᖅᑎᓪᓗᒋᓚ ᓴᓂᕐᕙᒃᓴᐅᕗᑦ ᐅᑭᐅᓚᒫᕐᒧᑦ.

Although you can collect paunnait to make tea all summer, the best time to collect it is the very late summer when its flowers are gone but its leaves are still green. At this time, it has the most flavour and medicinal strength. Once collected, the leaves can be saved and used for tea all winter.

ᐊᑲᐅᓯᓴᐅᑕᐅᓂᐊᖅᑎᓪᓗᒋᓪᓚ, ᐸᐅᓐᓇᐃᑦ ᑎᖅᑎᑦᑐᒥᒃᑖᑦᑎᓪᓗᒋᑦ ᐃᒪᓐᓚ ᐃᓱᓪᓚᕆᓚᖅᑎᓪᓗᒍ. ᑖᓐᓇ ᐊᑲᐅᓯᖀᓪᓕᖅᑎᑦᑎᔪᓐᓇᖅᑐᖅ ᓈᕐᓗᑦᑐᓂᒃ ᐊᔾᔩᙱᑦᑐᓂᒃ. ᑏᖖᐊᖑᓪᓗᑎ ᓈᕐᓗᑦᑐᓄᑦ ᐊᓇᕈᓐᓇᐃᓕᓯᒪᔪᓄᑦ ᐊᑐᖅᑕᐅᒐᔪᑦᑐᖅ. ᐅᖃᐅᔾᔨᑦ ᐸᐅᓇᐅᑉ ᓂᕆᑐᐃᓐᓇᕐᓗᒋᑦ ᓈᖖᔪᒧᑦ ᐊᑲᐅᓰᔭᑕᐅᔪᓐᓇᕐᒥᔪᑦ.

For medicinal uses, boil the mixture until it is very dark. This will produce a tea that can help with many ailments, especially those related to the stomach. This tea is commonly known to help with constipation. Simply eating a handful of the leaves will also help an upset stomach.

ᐸᐅᓐᓇᐃᑦ ᖃᐅᔨᒪᔭᐅᕆᕙᑦ ᓂᐊᖅᑯᖖᔪᓄᑦ ᐊᑐᖅᑕᐅᔪᓐᓇᕐᓂᖏᑦ. ᓂᐊᕐᑯᖅᓯᐅᑕᐅᓂᐊᖅᑎᓪᓗᒍᓚ ᐸᐅᓇᖅ ᑎᖅᑎᑎᓪᓗᒍ ᐃᓱᓪᓚᕆᓚᕋᓱᓐᓂᖓᓂ, ᕿᓐᓂᖅᑑᓗᓂᓗ. ᑎᓴᐃᖅᑎᓚᐅᕐᓗᒍ, ᐆᓇᕋᓗᐊᕐᓗᓂ (ᐅᓇᓗᐊᙱᓱᖓᖅᐸ ᐊᑐᕐᓂᐊᖅᑐᒧᑦ), ᖃᓪᓗᓈᖅᑕᖅ ᒪᓴᒃᑎᕐᓗᒍ, ᑕᒪᓐᓇ ᖃᓪᓗᓈᖅᑕ ᐃᓅᑉ ᓂᐊᕐᑯᖓᓄᑦ ᕿᓚᓪᓗᒍ. ᐊᑲᐅᓯᓴᐅᑎᖓ ᐱᕈᓯᐊᑦ ᐅᕝᕕᓂᒃᑯᑦ ᑎᒥᨀᖅᐸᓪᓚᐊᓂᐊᖅᐳᖅ.ᐆᓇᕐᓂᖓ ᑏ ᐃᓄᒻᒥᑦ ᐅᖅᑰᑎᑦᑎᓗᓂ ᐋᓐᓂᐊᒥᑦ ᐊᓂᐊᑎᑦᑎᔪᓐᓇᖅᑐᖅ ᐊᐅᒃᑲᕐᓂᒃᑯᑦ.

ᐸᐅᓐᓇᐃᑦ ᐱᕈᕐᑐᔪᑦ ᓯᐅᕋᕐᒥᑦ ᓱᕋᑦᑕᐅᓯᒪᔪᓂᑦ, ᓲᕐᓗ ᐊᖅᑯᑎᑯᑖᑦ ᑭᓪᓕᙱᓐᓂᒃ, ᐃᓪᓗᓕᐅᕐᕕᐅᔪᐃᑦ ᓴᓂᐊᓂᑦ, ᐊᓄᕞᒃᑲᕐᓂᓗ ᓄᓇᔭᒥᑦ. ᐊᔾᔨᓕᐅᕆᔭᕝᕕᓂᖅ ᓃᑦ ᑯᕆᓴᑕᕗᕐ.

Paunnait grow in places where the soil has been disturbed, such as along roadsides, near construction sites, and on windswept ridges of the tundra.

Paunnait are known to help with very bad headaches. To use paunnait for a headache, boil a very dark and thick tea. When the tea is still hot (as hot as the person can tolerate), soak a cloth in the tea. Wrap the soaked cloth around the person's head. The nutrients in the tea will be absorbed through the skin. The sweating caused by the warmth of the tea will also help.

ᐸᐅᓐᓇᐃᑦ ᐅᖅᑲᐅᔭᖏᑦ ᒪᑦᑐᑎᐅᖅᑲᑦᑕᕐᒥᔪᑦ ᑭᓚᖅᓯᒪᔪᒧᑦ, ᑭᑦᑐᕆᐊᖅᑕᐅᓯᒪᔪᒧᑦ ᐊᒻᒪᓗ ᐊᓯᖏᓐᓄᑦ ᐅᕙᓂᒻᒧᑦ ᑐᖄᖓᔪᑦ ᐊᑲᐅᙱᑕᐅᕿᑎᑦ. ᐅᐊᓴᑕᐅᖅᑎᑦᑐᒋᑦ ᐅᖅᑲᐅᔭᖏᑦ ᑭᑦᑕᖅᒧᑦ ᐃᑕᓗᒍ. ᓄᐊᖃᑖᖅᓯᒪᔪᑦ ᐅᖅᑲᐅᔭᐃᑦ ᑕᐃᒫᒃ ᐊᑐᖃᒃᓴᐅᔪᑦ, ᑭᓯᐊᓂ ᓄᐊᓯᒪᔪᑦ ᐅᑭᐅᓚᒫᖅᒧᑦ ᐊᑐᐃᓐᓇᓱᒃᓴᐅᓯᒍᑦ.

The leaves can also be used as small bandages to heal cuts, bug bites, or other skin irritations. After rinsing a leaf, place it on the cut or irritation. Fresh leaves can be used this way, but you can also save the leaves and use them as small bandages in the winter.

ᐸᐅᓐᓇᐃᑦ ᐃᓗᓕᐅᖓᔪᒥᒃ, ᔪᓚᐃ ᖅᑭᑎᖓᓂ ᐃᖅᑲᓗᖕᓂᑦ.

Paunnait along a ditch, mid-July in Iqaluit.

ᐸᐅᓐᓇᐃᑦ ᐊᑯᓪᓗᒋᑦ

Paunnait Alu

ᐊᑑᑎᖅᑲᑦᑎᐊᖅᑐᖅ ᓂᕿᑦᑎᐊᕙᐅᓪᓗᓂᓗ, ᓂᕆᔭᒃᓴᐅᒐᑦ ᐆᔪᖅᑐᕐᓗᓂ ᑐᒃᑐᕕᓂᕐᒥ ᓘᓐᓃᑦ.

- 5 ᐸᐅᓐᓇᐃᑦ ᐅᖅᑲᐅᔭᖏᑦ
- 2 ᐊᓘᑎᑯᔅᓯᒃ ᓇᑦᑎᐅᑉ ᐊᐅᖓ • 1 ᐃᕐᖑᓯᖅ ᓇᑦᑎᐅᑉ ᐅᖅᓱᖓ

ᓄᐊᑦᑎᓚᐅᕐᓗᑎᑦ ᑕᓪᓕᒪᓂᒃ ᐸᐅᓐᓇᓂᒃ, ᐅᖅᑲᐅᔭᖏᑦ, ᐊᒡᒐᖕᓄᑦ ᓱᕐᑦᑎᕐᓗᒋᑦ. ᐃᖑᓚᐅᒻᒨᕐᓗᒋᑦ ᒥᑭᓗᐊᖏᑦᑐᒧᑦ, ᓇᑦᑎᐅᑉ ᐊᐅᖓ ᐃᓚᓕᐅᑎᓗᒍ ᐃᖑᓚᕐᓗᒍ. ᐃᖑᓚᖅᓯᓐᓈᖅ ᓇᑦᑏᑉ ᐅᖅᓱᖓ ᐃᓚᓕᐅᑎᕙᓪᓕᐊᓗᒍ, ᐃᖑᓚᑦᑎᐊᕐᓗᒍ.

A nutritious side dish that can be served with caribou or other meat.

- Leaves of 5 paunnait
- 2 tablespoons of seal blood • 1 cup of seal fat

To collect the leaves of five paunnait, roll each stem between your hands until the leaves fall off. Gather the leaves in a bowl and stir in two tablespoons of seal blood. Next, gradually add one cup of fat, mixing very well.

ᖁᖑᓖᑦ

Qunguliit / Mountain Sorrel

ᖁᖑᓖᑦ ᑲᓇᑕᒥ ᓇᒥᑐᐃᓐᓇᓗᒃ ᐱᕈᕐᖑᔪᑦ, ᓄᓇᕗᓕᒫᐸᓗᒻᒥᓗ. ᐅᖅᑰᔭᖏᑦ ᑐᖑᔪᖅᑕᐅᓪᓗᑎᒃ, ᑕᖅᑐᙵᒍᐊᑯᓘᔭᖅᑐᑦ ᑲᑎᒻᒪᓪᓗᑎᑦ ᐱᕈᓯᑦ ᐊᒻᒪᓗ, ᑕᑭᔪᑯᑖᑯᓗᓐᓂᑦ ᐊᐅᐸᓪᓚᓇᑦᑐᓂᒃ ᓇᑯᖅᑲᖅᑐᑎᒃ. ᐅᖅᑰᔭᖏᑦ ᐊᖏᓂᖅᑖᓰᑦ 5-7 ᓯᐊᓐᑎᒦᑐᐸᓗᓐᓂᑦ. ᓇᑲᖏᑦᑕ ᐊᖏᓂᖅᓴᒻᒪᕆᐅᕚᑦᑐᑦ, ᑕᑭᓂᖅᑖᕈᓐᓇᖅᑐᑎᒃ 10-20 ᓯᐊᓐᑎᒦᑐᓂᑦ. ᖄᓂᕐᓂᐊᕈᕕᐅᑦᑕ, ᐊᐅᐸᕐᓂᖓ ᓇᑲᖓ ᓇᓂᓯᔾᔪᑎᒋᔪᓐᓇᖅᑕᐃᑦ.

Qunguliit grows in many places throughout Canada, including most parts of Nunavut. It has green, kidney-shaped leaves that grow in clumps and tall, dark red flower stalks that grow up from them. The leaves (known specifically as qunguliit) grow between 5 cm to 7 cm high. The flower stalks (known specifically as *nakait*) are much taller, growing 10 cm to 20 cm high.

ᖁᖑᓖᑦ ᒪᒪᕐᓂᖅᐹᑦᑎᐊᖑᕗᑦ ᓂᕆᔭᒃᓴᓂᑦ ᐱᕈᖅᑐᕕᓂᕐᓂᑦ. ᐅᖅᑰᔭᖏᑦ ᓇᑲᖏᓪᓗ ᐃᒃᓯᖅᑳᑎᐊᖅᑐᑎᑦ ᓲᕐᓇᕈᔪᒃᑐᑯᓘᓪᓗᑎᒃ. ᓂᕆᓂᐊᕐᓗᒋᑦ ᖁᖑᓖᑦ, ᓄᐊᑕᑎᑦ ᐊᒃᓴᓕᓪᓗᒋᑦ ᑲᑎᑦᑎᐊᕐᓗᒋᑦ, ᐱᔭᓐᓇᖅᑕᑎᑦ. ᑕᐃᒫ ᓲᕐᓇᖅᑐᑦ ᖅᑲᒃᑭᓐᓇᕐᓂᖅᓴᐃᓪᓗ ᑲᑎᑦᑎᐊᕐᓗᒋᑦ ᒪᒪᖅᑐᑯᓗᓐᓂᖅᑲᑦᑕᖅᑐᑦ. ᖁᖑᓖᑦ ᓂᕿᑦᑎᐊᕚᓗᐃᑦ ᐊᒻᒪ ᐅᖅᑰᑕᐅᓰᖑᓪᓗᑎᑦ ᕙᐃᑕᒪ ᑕ-ᖅᑲᓇᐊᖓ.

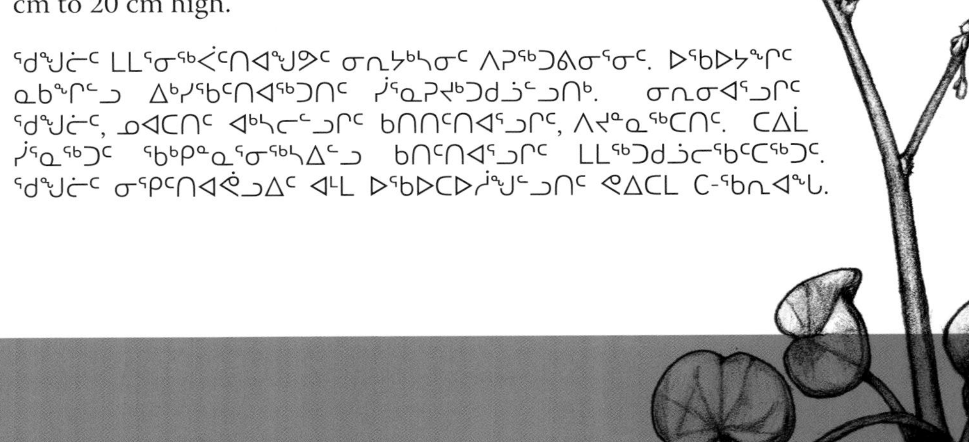

To find this plant, just look for its red flower stalks. This plant is a treat among all the edible plants. The juicy leaves and stems are tangy with a touch of sweetness. To enjoy this plant, rub a handful of the green leaves and red stems between your palms, squishing them together before you eat them. This combines the sweet and sour juices of the plant to make the most delicious flavour. The leaves and stems are very nutritious and is said to be a good source of vitamin C.

ᓯᓈᓇ ᑲᓪᓗᒃ, ᐅᖃᓕᒫᒐᓕᐅᖅᑑᑉ ᐱᖃᑎᖓ, ᐆᒃᑐᕈᓐᓇᕐᓂᕋᐃᒻᒥᔪᖅ ᐊᒡᒍᒐᓛᓪᓗᒋᑦ ᐅᖃᐅᔭᖏᑦ ᓇᑲᖏᓪᓗ ᑲᔪᕐᒥᑦ ᓱᑲᕐᒥᑦ ᐃᓚᓗᒍ ᓂᕆᓗᓂᑦ ᐃᓚᕐᑎᐊᕙᓘᓂᕋᐃᓪᓗᓂ. ᖁᖑᓖᑦ ᐱᕈᖅᑐᕕᓂᕐᓄᑦ ᑐᖑᔪᖅᑕᓄᑦ ᐃᓚᓕᐅᔾᔭᒃᓴᐅᒻᒥᔪᑦ, ᓲᕐᓗ ᐸᐅᓐᓇᓄᑦ (p. 44) ᐊᒻᒪ ᓯᐅᕋᐅᑉ ᐅᖃᐅᔾᔨᖏᓐᓄᑦ (p. 84)

Celina Kalluk, a friend of the authors, suggests chopping the leaves and flower stalks into small pieces and tossing them with brown sugar to make a delicious side dish. Qunguliit can also be combined with other greens, such as *paunnait* (dwarf fireweed; p. 44) and *siuraup uqaujangit* (seaside bluebells; p. 84).

ᖁᖑᓖᑦ ᐊᐅᔭᓕᒫᖅ ᓂᕆᔭᒃᓴᐅᔪᑦ, ᑭᓯᐊᓂᓕ ᒪᒪᓛᖑᖅᑲᑦᑕᖅᑐᑦ ᐊᐅᔭᐅᑉ ᐃᓱᐸᓗᐊᓂᑦ.

Qunguliit can be eaten all summer long, but it is most nutritious and tasty toward the end of summer.

ᐅᕐᑲᐅᔭᖅᑎᑦ ᓇᑲᖅᑎᓪᓗ ᐆᑕᒃᓴᐅᒻᒥᔪᑦ. ᐊᕐᑭᓪᓕᕙᔭᖅᑐᑦ ᑲᔪᑭᔫᓕᕐᓗᑎᓪᓗ ᐊᑯᓂᐅᖅᑎᑦᑐᖅ ᑎᖅᑎᑦᑐᒦᓚᐅᖅᑎᓪᓗᑎᑦ. ᐅᕐᑲᐅᔭᑦ ᓇᑲᓪᓗ ᐆᓯᒪᓕᖅᑎᓪᓗᑎᑦ ᓂᕆᑐᐃᓐᓇᖃᒃᓴᐃᑦ ᐅᖕᖏᓘᓐᓃᑦ ᓂᕆᓂᐊᖅᑕᓐᓄᑦ ᐃᓚᓕᐅᑎᓗᑎᑦ. ᓴᙱᓗᐊᖅᑎᑦᑐ ᑎᐱᖓ ᒪᒪᖅᑐᓂᓗ. ᖃᔪᖓᓗ ᓂᓪᓕᔅᓯᐊᕆᓚᐅᕐᓗᒍ ᐃᒥᖅᑕᐅᔪᓐᓇᖅᓯᓗᓂ. ᑖᓐᓇ ᖃᔪᖓ ᓄᑕᕋᕐᓄᑦ ᐃᒻᒧᒃᓴᖅᑲᖅᑎᑦᑐᓄᑦ ᐱᔭᐅᔪᓐᓇᖅᑐᖅ. ᐊᑲᐅᓯᓴᐅᑎᐅᑎᓪᓗᓂᑦ ᐃᒥᖅᑕᐅᔪᓐᓇᕐᒥᔪᑦ ᓂᕆᔪᒻᒪᖅᑎᑦᑐᐊᓘᓕᖅᑐᓄᑦ.

The leaves and flower stalks can also be cooked. They will become tender and light brown when they are boiled briefly. The boiled leaves and stalks can be eaten alone or served as part of a meal. Their flavour will

ᐋᓚᓯᐅᑉ ᐊᒃᓴᓕᑦᑕᖅᑎᑦ ᐊᒡᒐᒥᓄᑦ ᕐᑯᖕᒍᓖᑦ ᐃᒃᓯᓛᑦ ᔫᕐᓇᓕᓈᓂᑦ ᑲᑎᑎᒐᓯᒃᑕᖅᑎᑦ.

Aalasi rubs a handful of *qunguliit* between her palms to combine the plant's sweet and sour juices.

be mild and pleasant. Also, the clear broth that is left over can be chilled and served as a refreshing drink. This drink may be given to children when milk is not available. It can also be used medicinally to revive a person who has lost their appetite.

ᖁᖑᓕᐅᑉ ᐊᒫᖏᑦ ᒪᒪᖅᑐᑯᓘᒻᒥᔪᑦ ᓂᕿᑦᑎᐊᕚᓪᓗᑎᒡᓗ. ᑭᓯᐊᓂ, ᐆᖃᓯᐅᑎᑦᑕᐃᓕᒃᑭᑦ ᐆᓯᒪᓪᓗᑎᑦ ᒪᒪᕈᓐᓂᓲᖑᒻᒪᑕ. ᓂᕆᓂᐊᕐᓗᒋᓕ, ᐲᔭᕐᓗᒋᑦ ᓂᕆᑐᐃᓐᓇᖁᔪᖅ.

The rhizomes (underground stems) are also tasty and very nutritious. But, do not boil them because they are not tasty when cooked. To eat them, just dig them up and enjoy.

ᖁᖑᓖᑦ ᐱᕈᓕᑐᐊᖅᐸᑕ ᓂᕆᔭᒃᓴᐃᑦ,ᑭᓯᐊᓂ ᒪᒪᑖᖑᒋᕗᑦ ᐊᐅᔭᒃᑯᑦ. ᑕᒫᓂᕈᔪᒃ ᑐᖑᑦᑐᕆᑦᑐᑯᓘᕚᑦᑐᑦ ᓂᖅᑭᕆᑦᑐᑎᒃ, ᐅᑭᐊᒃᓴᙳᑐᐊᖅᐸᑦ ᑕᕙ ᐊᐅᐸᕈᔪᒃᓯᓂᐊᕆᓪᓗᑎᑦ.

The leaves and flower stalks can be eaten as soon as they have grown, but are most nutritious and flavourful in the late summer. At this time, the leaves will still be green and fleshy, but soon they will turn bright red as the season turns to early autumn.

ᐋᓚᓯ ᓇᐃᒪᔪᖅ ᒪᒪᑦᑐᖅ ᖁᖑᓕᓂᑦ. ᖁᖑᓖᑦ ᐃᓅᓕᓴᐅᑎᖃᑦᑎᐊᖅᑐᑦ ᕙᐃᑕᒪᓐ ᑕᓂᑦ.

Aalasi takes in the delicious smell of a handful of *qunguliit*. Qunguliit is high in vitamin C.

ᖁᖑᓕᐅᑉ ᖃᔪᖓ (ᑎᓴᐃᖓᔪᖅ ᑏᖖᒍᐊᖅ)
Qunguliit Iced Tea

- ᓄᐊᓯᒪᔪᑦ ᐊᒡᒐᓐᓃᑦᑐᐊᖅᑐᑦ ᖁᖑᓖᑦ (ᐊᐃᕋᖅᑲᖕᒥᑦᑐᒥᑦ)
- 5 ᐆᕙᓘᓐᓃᑦ 6 ᐃᕐᖑᓯᑦ ᐃᒪᖅ
- ᓂᓚ

ᑎᖅᑎᑎᓪᓗᒋᑦ ᖁᖑᓖᑦ ᑲᔪᕈᔪᕈᕋᓱᓐᓂᖕᒋᓐᓂᑦ. ᐲᔭᕐᑐᒋᑦ ᖁᖑᓖᑦ (ᓂᕆᓂᐊᑐᐃᓐᓇᕐᑐᒋᑦ) ᐊᒻᒪ ᑎᓴᐃᖅᓯᐊᓇᓗᒍ ᖃᔪᖓ. ᖃᔪᖅ ᓂᓚᓂᑦ ᐃᓗᑦᓚᖅᓗᒍ ᐱᔭᒃᓴᑦᑎᐊᖑᒻᒥᔪᖅ. ᐅᐊᑦᑎᕈᓘᓐᓃᑦ ᐱᖕᒑᖅᓂᐊᖅᓗᒍ ᖁᐊᒃᑯᕕᒻᒦᑎᑕᒃᓴᐅᒥᑦᑐᓂ.

- Several handfuls of fresh qunguliit (with roots removed)
- 5 or 6 cups of water
- Ice

Boil the qunguliit until they turn light brown. Remove the qunguliit (set aside to eat) and then allow the cooking water to cool. Serve the cooking water over ice and enjoy. It can also be refrigerated and saved for later.

ᓴᐸᖓᕋᓛᙳᐊᑦ ᑐᖅᑕᐃᓪᓗ
Sapangaralaannguat Tuqtaillu / Alpine Bistort

ᓴᐸᖓᕋᓛᙳᐊᑦ, ᐃᓄᒃᑎᑐᑦ ᑐᑭᖃᖅᑐᑦ "ᓴᐸᖓᐅᔮᖅᑐᑦ", ᑖᓐᓇ ᓄᓇᕋᐅᑉ ᓴᖅᑭᔮᕐᓂᖓᓂ ᐅᖃᐅᓯᓕ. ᑐᖅᑕᐃᑦ ᐅᖃᐅᓯᖃᖅᐳᖅ ᓄᓇᕋᐅᑉ ᐃᓚᖓᓂ (ᓇᑲᖓᑕ ᐃᓚᖓ ᓄᓇᑉ ᐊᑖᓂᑦᑐᖅ). ᐅᖃᓕᒫᒐᕐᒥᒃ ᑖᓐᓇ ᓇᓗᓇᐃᖅᓯᒪᔭᕗᑦ ᓴᐸᖓᕋᓛᙳᐊᑦ ᑐᖅᑕᐃᓪᓗ ("ᓴᐸᖓᕋᓛᙳᐊᑦ ᐊᒻᒪᓗ ᑐᖅᑕᐃᑦ") ᐱᔾᔪᑎᒋᓪᓗᒍ ᑕᒪᕐᒦᒃ ᐅᖃᐅᓯᐅᑲᑕᓐᓂᐊᕋᒦᒃ ᐊᑐᓂ.

Sapangaralaannguat, which means "miniature imitation beads" in Inuktitut, refers to the above-ground part of this plant. *Tuqtait* refers to its rhizomes (underground stems). We have called this plant *sapangaralaannguat tuqtaillu* ("sapangaralaannguat and tuqtait") because both parts of the plant are discussed here together.

ᓴᐸᖓᕋᓛᙳᐊᑦ ᑕᑭᔪᑯᑖᑯᓘᕗᑦ, ᖃᑯᖅᑕᓂᒃ ᐱᕈᖅᓯᐊᖃᖅᑐᑎᒃ ᖄᖓᒍᑦ. ᐱᕈᖅᓯᐊᖑᖏᑦ ᒪᑉᐱᓚᐅᖅᑎᓐᓇᒋᑦ, ᓴᐸᖓᑯᓘᔭᖃᑦᑕᖅᑐᑦ, ᐊᐅᐸᖅᑐᑎᒃ. ᑐᖑᔪᖅᑕᐅᓪᓗᑎᒃ ᐅᖃᐅᔭᖏᑦ ,ᐊᒥᒃᑐᑯᑖᑯᓘᑦᓗᑎᒃ,ᕿᑦᓕᕈᔪᒃᑐᑎᒃ. ᓇᑲᖓᑕ ᐊᑖᓂ ᐱᕈᖅᑲᑦᑕᖅᑐᑦ. ᓴᐸᖓᕋᓛᙳᐊᑦ ᑕᑭᓂᖅᑲᓪᓚᕈᓐᓇᖅᑐᑦ 25 ᓯᐊᓐᑎᒦᑐᒥᑦ.

Sapangaralaannguat, the above ground part of this plant, has tall, green stalks with little white flowers at the top. Before the flowers open, they look like small, red beads. The green leaves are long, narrow, and shiny. They grow out from the base of the stalk. Sapangaralaannguat can grow up to 25 cm high.

ᑎᓯᔪᑯᓗᐃᑦ “ᓴᐸᖓᙳᐊᑦ” ᓂᕆᔭᒃᓴᑦᑎᐊᖑᔪᑦ ᐱᕈᖅᓯᐊᙳᓪᓚᐅᖏᓐᓂᖏᓐᓂᑦ. ᐅᖅᑲᐅᔾᖏᓪᓗ ᓂᕆᔭᒃᓴᐅᒋᓪᓗᑎᑦ, ᑭᓯᐊᓂ ᓂᕆᔭᐅᒐᓪᓚᖏᑦᑐᑦ ᓂᕆᔪᒥᓇᓗᐊᖏᒻᒪᑕ.

The hard “beads” are enjoyable to eat before they turn into flowers. The leaves are also edible, but they are only eaten if necessary because they have an unpleasant texture.

ᐋᓚᓯ ᐃᖅᑲᐅᒪᔪᖅ ᓂᐱᐊᖅᓯᐊᖑᓪᓗᓂ “ᓴᐸᖓᙳᐊᑦ” ᐱᙳᐊᕆᐼᑦᑐᓂᒋᑦ. ᓄᐊᑕᐅᔪᓇᖅᑐᑦ ᐊᐃᑦ ᐱᕆᓯᒪᓗᒍ ᑕᒪᐅᖓ ᐱᕆᓐᓂᖓᓄᑦ, ᖃᖓᑐᐃᓐᓇ ᐅᓪᓘᑉ ᐃᓗᐊᓂ ᓂᕐᒪᑲᑕᓐᓂᐊᕐᓗᒋᑦ. ᓴᐸᖓᕋᓛᙳᐊᑦᓄᐊᓪᓗᒋᑦᖃᓪᓗᓅᖅᑕᒎᑦᕿᓚᓪᓗᒋᑦ “ᑎᒎᒥᐊᒐᖅ” ᑎᒎᒥᐊᖅᑲᑦᑕᓕᐊᕆᔭᒃᓴᐅᒥᔪᑦ. ᐊᒐᓐᓂᑦ ᑎᒎᓪᓚᖅᐸᓪᓗᒍ.

Aalasi recalls playing with the beads as a young girl. They can be collected in the pouch of a sleeve folded up at the wrist, and eaten throughout the day. The beads can also be collected and tied into a small sack to create a soothing “stress ball.” Roll the sack in the palm of your hand.

ᓴᐸᖓᕋᓛᙳᐊᑦ (“ᓴᐸᖓᐅᔮᖅᑐᑦ ᒥᑭᑦᑐᑯᓗᐃᑦ) ᐊᑎᖅᑲᐅᑎᖅᑲᖅᑐᑦ ᒥᑭᑦᑐᑯᓗᐃᑦ, ᐊᐅᐸᖅᑐᑦ “ᓴᐸᖓᙳᐊᖏᓐᓄᑦ” ᐱᕈᖅᓯᐊᙳᕆᑦ ᐊᐅᔭᐅᑉ ᕿᑎᐸᓗᐊᓂᑦ.

Sapangaralaannguat (“miniature imitation beads”) is named for its tiny, red “beads” that turn into flowers mid-season.

ᑐᖅᑕᐃᑦ, ᓇᑲᖕᒐᑕ ᐃᓚᖕᒐ, ᒪᒪᖅᑐᒻᒪᕆᐅᔪᑦ ᓂᖅᑭᑦᑎᐊᕙᐅᓪᓗᑎᓪᓗ. ᐃᓗᒥᑦ ᓴᓗᒻᒪᐃᑦᑎᐊᕈᓐᓇᖅᒥᔪᑦ.
ᑎᓯᓂᖕᒐ ᑎᐱᖕᒐᓗ ᖃᖅᑯᐊᖅᑲᑎᑑᐸᓗᒃᑐᑦ. ᑐᖅᑕᐃᑦ ᐳᑐᐃᓐᓇᖃᒃᓴᐅᕗᑦ.

Tuqtait, the rhizomes, are very tasty and nutritious. They also cleanse the digestive system. Their texture and taste are similar to almonds. Tuqtait can be enjoyed fresh from the ground. Use a small shovel or your fingers to dig them up. Detach them from the rest of the plant and clean off the small white stems with your fingernail.

ᐆᓯᒪᓪᓗᑎᓪᓚ, ᑐᖅᑕᐃᑦ ᐊᖅᑭᑦᑐᑯᓘᕗᑦ, ᓵᖅᓗᑦᓚ ᒑᑏᓯᖃᓛᑎᑐᑦ. ᐆᓯᒪᓗᑎᑦ ᓂᓇᓂᐊᕈᕕᒋᑦ, ᑎᖅᑎᑎᖅᑯᔪᑦ 5-10 ᒪᓇᐸᓗᒻᒥᑦ, ᐊᖅᑭᑦᓯᓇᓱᓐᓂᖕᒋᓐᓂᑦ.

When cooked, tuqtait are mild and soft, like very tiny potatoes. To enjoy them cooked, boil them until they are tender, about five to ten minutes.

ᑐᖅᑕᐃᑦᐊᒡᒍᒐᓛᑦᑐᒥᑦᓂᖅᑭᓄᑦᐃᓚᒃᓴᑦᑎᐊᕙᐅᒋᕗᑦ. ᐊᒡᒍᒐᓛᒃᓯᒪᔪᑦ ᐸᓚᐆᒐᓕᐊᒍᑦ ᐃᓚᓕᐅᑎᔭᒃᓴᐅᒋᕗᑦ ᐆᓚᐅᖕᒋᓐᓂᖅᓂᑦ ᐸᓚᐆᒐᖅ. ᐃᓚᓕᐅᑎᒐᒃᓴᐅᒋᕗᑦ

ᐋᓚᓯ ᓄᐊᑦᑎᑎᓪᓗᒍ ᑐᖅᑕᓂᑦ (ᐊᒫᖕᒋᓐᓂᒃ), ᖃᓗᖁᖅᑐᓂᒋᑦ ᐲᔭᖅᑐᓂᒋᑦ ᓴᐸᖕᒐᖁᓛᙳᐊᓂᑦ (ᓄᓇᐅᑉ ᖄᖕᒐᓃᓐᓂᖕᒐ ᓄᓇᖁᐅᑉ).

Aalasi harvests *tuqtait* (the rhizomes), digging them up and detaching them from the *sapangaralaannguat* (the above-ground part of the plant).

ᑲᑎᑕᐹᓇᑦ ᐅᖅᑲᐅᔭᓇᑦ ᑐᖑᔪᖅᑕᓇᑦ ᐆᒃᑑᑎᒋᔪᒍ, ᑲᑎᓯᒪᔪᑦ ᖁᖑᓖᑦ (p. 52) ᐸᐅᓐᓇᐃᑦᑐ (p. 44) ᒪᒪᖅᑐᐊᓘᕗᑦ ᑐᖅᑕᓂᒃ ᐃᓚᓯᒪᓪᓗᒋᑦ. ᑐᖅᑕᐃᑦ ᑎᓯᓂᖅᓴᒻᒪᕆᐅᖅᑯᒍᕕᒋᑦ ᒪᒪᕐᓂᖅᓴᐅᔪᓂᔪ, ᐅᓪᓗᓕᒫᖅ ᓂᓪᓕᓇᖅᑐᒥᒃ ᐃᒫᓃᑎᖅᑯᔪᖅ ᓂᕆᓚᐅᖏᓐᓂᕐᓂᑦ.

Tuqtait can also be used as a seasoning for other foods by chopping them into small pieces. The pieces can be added to the batter of bannock before baking or frying it. They can also be added to greens. For example, a mixture of *qunguliit* (mountain sorrel; p. 52) and *paunnait* (dwarf fireweed; p. 44) is delicious topped with tuqtait. To make tuqtait even crunchier, soak them in cold water for a day before eating.

ᐋᓚᓯ ᑐᖅᑕᒥᑦ ᖁᐱᓯᑎᓪᓗᒍ. ᑐᖅᑕᐃᑦ ᖄᖅᑯᐊᒐᖅᑎᑐᑦ ᒪᒪᖅᑎᕆᔪᑦ ᑎᓯᕿᔫᒃᑐᑯᓘᓪᓗᑎᓪᓗ.

Aalasi breaks open *tuqtait*. Tuqtait have a mild almond flavour and an enjoyable crunch.

ᑐᖅᑕᐃᑦ ᓂᕿᑦᑎᐊᕚᓗᐃᑦ. ᐆᓯᒪᖕᒋᓪᓗᑎᒃ ᓂᕆᔭᒃᓴᐃᑦ ᖄᖅᑯᐊᒐᖅᑎᑐᑦ, ᐆᑦᑎᓗᒋᓪᓘᓐᓃᑦ, ᓲᕐᓗ ᑖᑎᓯᕋᓛᑎᑐᑦ.

Tuqtait are very nutritious. They can be enjoyed raw, like nuts, or cooked, like tiny potatoes.

ᑐᖅᑕᖅ ᐃᖃᓗᕕᓂᓕᒃ ᐸᐅᕐᖕᒐᐃᓪᓗ

Char Salad with Tuqtait and Paurngait (Crowberries)

- 6 ᑐᖅᑕᐃᑦ, ᐊᒡᒍᒐᓛᔅᓯᒪᔪᑦ
- 1 ᓇᑉᐸᖅ ᐃᖃᓗᕕᓂᖅ (2-3 ᐃᓐᑦᓯᔅ ᓯᑕᓐᓂᓕᒃ), ᓴᐅᓃᔭᖅᓯᒪᔪᖅ ᐊᒥᖅᑲᕋᓂᓗ ᐆᔪᖅ
- 1 ᐃᕐᖕᒍᓯᖅ ᐸᐅᕐᖕᒐᐃᑦ (p. 66) ᐊᓯᖕᒋᑦᑑᓐᓂᑦ ᐸᐅᕐᖕᒐᑐᐃᓐᓇᐃᑦ
- 2 ᐊᔫᑎᑯᔅᓯᒃ ᐄᔭᖅᓂᕈᑎ, ᐱᑕᖅᑲᕈᓂ ᑭᓯᐊᓂ (ᐊᓯᖕᒋᑦᑑᓐᓃᑦ ᐅᖅᓱᑐᐃᓐᓇᖅ, ᓇᑦᑎᐅᑦᑑᓐᓃᑦ ᐅᖅᓱᖕᒐᓂ)

ᐃᖃᓗᕕᓂᖅ ᓵᒨᕐᓗᒍ ᐃᖕᒍᓚᕐᓗᒍ ᓱᖁᕐᑎᕐᓗᓂ. ᐃᓚᓕᐅᑎᓗᒋᑦ ᑐᖅᑕᐃᑦ ᐄᔭᖅᓂᕈᑎᓗ ᐊᒻᒪ ᐃᖕᒍᓚᕐᑎᐊᕐᓗᒋᑦ. ᐸᐅᕐᓇᐃᑦ ᐃᓚᓕᐅᑎᑎᕐᒥᓗᒋᑦ, ᖃᒐᖅᑕᐃᓕᒪᓗᒋᑦ ᐃᖕᒍᓚᒃᑲᓐᓂᕐᓗᒋᑦ. ᐆᓇᕈᔪᓐᓂᖕᒐᓂᑦ ᓂᕆᔭᐅᔭᕆᐊᓕᒃ.

- 6 tuqtait, chopped into small pieces
- 1 section of a char (about 2 to 3 inches thick), boiled with all skin and bones removed
- 1 cup of paurngait (crowberries; p. 66) or other berries
- 2 tablespoons of mayonnaise, if available (or other fat, such as seal fat)

Place the boiled char in a bowl and break it into flakes. Add the tuqtait and mayonnaise and mix well. Add the berries, mixing gently so they are not squished. Serve warm.

ᐅᑭᐊᒃᓵᒃᑯᑦ, ᑲᓪᓚᖁᑎᑦ ᖃᖅᑳᔮᓂᑦ ᖃᐅᒻᒪᐃᓯᒪᔪᑦ ᐃᖃᓗᐃᑦ ᓯᓚᑎᑦᑎᐊᖓᓂᑦ ᐃᒪᐅᑉ ᓴᓂᐊᓂᑦ. ᑲᓪᓚᖁᑏᑦ ᑏᑦᑎᐊᕚᓗᐃᑦ ᒪᒪᖅᑐᐊᓗᐃᑦ.

In autumn dusk, *kallaqutit* (bearberry leaves) light up a hillside overlooking Frobisher Bay near Iqaluit. Kallaqutit make a deliciously tangy tea.

ᐸᐅᕐᖓᐃᑦ ᐊᔾᔨᒌᖏᑦᑐᑦ
Paurngait Ajjigiingittut / Berry Plants

ᐊᐅᔭᐅᑉ ᕿᑎᓪᓚᓂᖓᓂᑦ ᓇᒥᑐᐃᓐᓈᓗᒃ ᐸᐅᕐᖓᑯᓗᖕᓂᑦ ᐅᑭᐅᖅᑕᖅᑐᖅ ᐱᕈᕐᕕᐅᓲᖑᒍᖅ. ᐱᖓᓲᒍᑦᑕ ᓇᑐᓇᐃᕐᓂᐊᖅᑕᒍᑦ ᐅᕙᓂ: ᐸᐅᕐᖓᐃᑦ ᐸᐅᕐᖓᕐᑯᑎᓪᓗ, ᑲᓪᓚᐃᑦ ᑲᓪᓚᖅᑯᑎᓪᓗ, ᐊᒻᒪᓗ ᑭᒍᑕᖏᕐᓇᑦ ᓇᕐᑯᑎᓗ. ᐸᐅᕐᖓᐃᑦ ᐊᔾᔨᒌᖏᑦᑐᑦ ᑖᒃᑯᐊᑦ ᐋᓚᓯᐅᑉ ᖃᐅᔨᒪᓛᓂᕙᐃᑦ.

In late summer, many regions of the Arctic are rich with juicy berries. Three different berry plants are described here: *paurngait paurngaqutillu* (crowberry), *kallait kallaqutillu* (bearberry), and *kigutangirnait naqutillu* (blueberry). These are the berry plants with which Aalasi is most familiar.

ᐸᐅᕐᖓᐃᑦ ᐸᐅᕐᖓᕐᑯᑎᓪᓗ

Paurngait Paurngaqutillu / Crowberry

ᑭᒍᑕᖏᕐᓇᐃᑦ ᓇᕐᑯᑎᓪᓗ

Kigutangirnait Naqutillu / Blueberry

ᑲᓪᓚᐃᑦ ᑲᓪᓚᕐᑯᑎᓪᓗ

Kallait Kallaqutillu / Bearberry

ᐸᐅᕐᖓᐃᑦ ᐸᐅᕐᖓᖁᑎᓪᓗ

Paurngait Paurngaqutillu / Crowberry

ᐃᓄᒃᑎᑐᑦ ᐸᐅᕐᖓᐃᑦ (ᐊᑕᐅᓯᐅᓪᓗᓂ: ᐸᐅᕐᖓᖅ) ᐸᐅᕐᖓᓂᑦ ᐱᕈᖅᑐᓂᑦ ᐅᖃᐅᓯᖃᖅᐳᖅ ᐊᒻᒪᓗ ᐸᐅᕐᖓᖁᑎᑦ ᐅᖃᐅᓴᖕᒋᓐᓂᑦ ᐱᕈᖅᕕᖕᒋᓐᓂᑦ ᐅᖃᐅᓯᖃᖅᑐᑎᑦ. ᐅᕙᓂᓕ ᑕᐃᑲᑕᑉᐸᒍ ᐸᐅᕐᖓᐃᑦ ᐸᐅᕐᖓᖁᑎᓪᓗ.

In Inuktitut, *paurngait* (singular: *paurngaq*) refers to the berries of this plant and *paurngaqutit* refers to the leaves. Here, we refer to this plant generally as *paurngait paurngaqutillu* ("paurngait and paurngaqutit").

ᐸᐅᕐᖓᐃᑦ ᐸᐅᕐᖓᕐᑯᑎᓪᓗ ᐱᑕᕐᑲᒐᔪᑦᑐᑦ ᓄᓇᓕᖕᓂᑦ ᓄᓇᕗᒻᒥᑦ (ᐱᑕᕐᑲᓗᐊᖕᒋᑦᑐᑦ ᑭᓯᐊᓂ ᕿᑎᕐᒥᐅᓂᑦ). ᐃᓕᓴᕆᔭᕐᓂᖅᐳᑦ ᐅᕐᑲᐅᔭᖕᒋᑦ ᑐᖑᑦᑐᕆᐅᔭᐃᓐᓇᕐᒪᑕ, ᒥᖅᑯᑎᐅᔮᖅᑐᑎᑦ ᖃᔫᓂᖕᒐᓂᒃ ᓇᑲᖕᒐᓃᖖᒑᖅᑐᑦ. ᓇᑲᖕᒋᑦ ᐱᕈᕐᑲᑦᑕᖅᑐᑦ ᓯᓕᑦᑐᑎᒃ ᐊᒻᒪᓗ ᓂᒐᕕᐅᖅᑲᔪᑰᓪᓗᑎᒃ ᐊᒻᒪ 15 ᓯᐊᓐᑎᒦᑐᒥᑦ ᐊᖏᓂᖅᓴᐅᓕᕈᓐᓇᖏᑦᑐᑦ. ᐃᔅᓯᓂᑦ ᐊᒥᐊᓕᓐᓂᑦ ᐱᕈᖅᓯᐊᑖᑲᐃᓐᓇᕐᑲᑦᑕᖅᑐᑦ ᑭᓯᐊᓂ ᕿᓚᒥᑯᓗᒃ (ᕿᓂᑲᐅᖅᑐᖕᒋᒃᑯᕕᒋᑦ, ᑕᑯᖖᒋᑕᐃᓐᓇᕐᓂᐊᖅᐸᑎᑦ!). ᐸᐅᕐᖓᐃᑦ ᐊᒻᒪᓗᑭᓵᖑᕗᑦ ᐃᒃᓯᕐᑲᑦᑎᐊᖅᑐᑎᑦ. ᐱᕈᓕᓵᖅᑐᑎᑦ ᑐᖑᔪᖅᑕᐅᕐᑲᑦᑕᖅᑐᑦ ᑕᕝᕙ ᕿᓐᓂᖅᓯᓕᓇᓪᓗᑎᑦ ᓂᓇᔭᒃᓴᐅᓕᖄᖕᒪᑕ ᐊᐅᔭᒃᑯᑦ.

Paurngait paurngaqutillu is common around many communities in Nunavut (although somewhat less common in the Kitikmeot). It is easily recognizable by its evergreen leaves, which look like short needles sticking out from the brown stems. The stems grow into thick tangles on the ground and are usually less than 15 cm high. The tiny, purple flowers appear very briefly in early spring (if you don't look quickly, you will miss them!). The paurngait are round and juicy. They are green when they first appear and then turn black and shiny when they are ripe in the late summer.

ᔪᓚᐃ ᐱᒋᐊᕐᓂᖕᒐᓂᑦ, ᐸᐅᕐᖓᐃᑦ ᑖᒃᑯᐊ ᐱᕈᓃᖕᒋᑦᑑᒐᓗᐊᑦ, ᑭᓯᐊᓂ ᒪᒪᖅᑐᑯᓘᓕᖅᑐᑦ ᐃᒃᓯᕐᑲᑦᑎᐊᖅᑐᑎᑦᓗ.

In early July, these *paurngait* are not ripe yet, but they already taste pleasant and refreshing.

ᓴᐅᓈᖕᒋᑦ ᖃᒃᑭᓐᓇᖅᑐᑦ, ᐸᐅᕐᖓᐃᑦ ᐳᕿᓚᔪᑦ ᓲᕐᓇᖅᑐᑯᓗᐃᑦ.

While the seeds are slightly bitter, the meat of ripe *paurngait* is tangy and sweet.

ᐸᐅᕐᖓᐃᑦ ᐸᐅᕐᖓᕐᑯᑎᓪᓗ ᖃᐅᔨᓴᕐᓂᖅᑐᑦ ᑐᖑᔫᓴᐃᓐᓇᕐᒪᑕ ᐅᕐᖃᐅᓴᕐᑳᖕᒋᑦ, ᑲᐱᓇᖅᑐᑯᓘᓴᖅᑐᑦ ᐱᕈᕐᕚᖕᒪᓃᖕᖔᖅᑐᑦ ᑲᔪᕐᒥᑦ.

Paurngait paurngaqutillu is easily recognizable by its evergreen leaves, which look like short needles sticking out from the brown stems.

ᐱᕈᓈᕐᑎᐊᖅᓯᒪᔪᑦ ᐸᐅᖕᒐᐃᑦ ᐸᐅᖕᒐᖅᑯᑎᓃᖅᑐᑦ ᐋᓚᓯᐅᑉ
ᐃᓪᓗᖕᒐᑕ ᖃᓂᐊᓂᑦ ᓰᑎᐱᓇᒻᒧᕋᑖᖅᑎᑦᑐᒧ.

A mat of ripe *paurngait paurngaqutillu* near Aalasi's house in early September.

ᑲᓪᓚᐃᑦ ᑲᓪᓚᖁᑎᓪᓗ
Kallait Kallaqutillu / Bearberry

ᐃᓄᒃᑎᑐᑦ, ᑲᓪᓚᐃᑦ (ᐊᑕᐅᓯᐅᓪᓗᓂ: ᑲᓪᓚᒃ) ᐸᐅᕐᖓᕐᒋᓐᓂᑦ ᐅᖃᐅᓯᖅᑲᖅᐳᖅ ᐊᒻᒪ ᑲᓪᓚᖁᑎᑦ ᐅᖃᐅᓴᕐᒋᓐᓂᑦ ᐱᕈᖏᕐᒋᓐᓂᑦ ᐅᖃᐅᓯᖅᑲᖅᑐᑎᑦ. ᐅᕙᓂᓕ ᑕᐃᑲᑕᓐᓂᐊᖅᑕᕗᑦ ᑲᓪᓚᐃᑦ ᑲᓪᓚᖁᑎᓪᓗ.

In Inuktitut, *kallait* (singular: *kallak*) refers to the fruit and *kallaqutit* refers to the leaves of the plant. Here, we refer to the plant generally as *kallait kallaqutillu* ("kallait and kallaqutit").

ᑲᓪᓚᐃᑦ ᑲᓪᓚᖁᑎᓪᓗ ᐱᑕᖃᒻᒪᕆᑦᑐᒻᒪᕇᑦ ᖀᖅᑕᓂ, ᑭᕙᓪᓕᕐᒥᓗ ᓄᓇᐃᑦ ᐃᓚᖏᓐᓂᑦ, ᐊᒻᒪᓗ ᕿᑎᕐᒥᐅᓂ. ᑲᓪᓚᐃᑦ ᑲᓪᓚᖁᑎᓗᑦ ᓇᓗᓇᖏᑦᑎᐊᖅᑐᑦ ᑕᒻᒪᐅᑕᐅᕙᓇᕆᑎᓪᓗ ᑭᒍᑕᖏᕐᓇᐃᑦ ᓇᖁᑎᖏᓐᓂᓪᓗ. ᐅᖃᐅᓴᖏᑕ ᑭᓪᓕᖏᑦ ᑐᑭᓕᐊᑦᑎᐊᖏᒻᒪᑕ (ᑭᒍᑕᖏᕐᓇᐃᑦ ᓇᖁᑎᓪᓗ ᑐᑭᓕᐊᑦᑎᐊᖅᑐᓂᑦ ᐅᖃᐅᓴᖃᕐᒪᑕ). ᐅᖃᐅᓴᖏᑦ ᑕᑯᔅᓴᐅᑎᑦᑎᐊᖅᑐᑦ

ᐱᕈᖅᓯᐊᕋᓛᑦ ᓄᕗᐊᓃᑦᑐᑦ ᑲᓪᓚᐃᑦ ᑲᓪᓚᖁᑏᓪᓗ ᓴᖅᑮᔭᑲᐃᓐᓇᑐᐃᓐᓇᖅᑲᑦᑕᖅᑐᑦ ᐊᐅᔭᐅᓚᓵᖅᑎᓪᓗᒍ. ᐅᕙᓂ ᐊᔾᔨᙳᐊᒥᑦ ᐅᖃᐅᓴᖏᑦ ᑐᙵᔪᖅᑐᑦ ᐊᕿᓂᓴᕐᓃᖅᑕᐅᔪᑦ ᑲᔪᕐᓂᑦ.

The tiny blossoms of *kallait kallaqutillu* appear briefly in the very early summer. In this picture, the new season's green leaves are crowded amongst the brown leaves of the previous season.

ᑲᓪᓚᐃᑦ ᑲᓪᓚᖁᑎᓪᓗ ᐱᕈᖅᐸᓪᓕᐊᔪᑦ ᓂᒐᕕᓯᒪᓪᓗᑎᑦ ᐃᕕᖕᓄᑦ, ᐊᓯᖏᓐᓄᓪᓗ ᐸᐅᖕᒪᓄᑦ, ᐊᒻᒪᓗ ᕿᔪᒃᑖᖅᐸᓄᑦ (ᑏᖕᒍᐊᒃᓴᐃᑦ). ᐅᑭᐊᒃᓴᓕᖅᑎᓪᓗᒍ, ᐊᔾᔨᒌᖏᑦᑐᑦ ᑲᓪᓚᐃᑦ ᐋᓚᓯᐅᑉ ᐃᓪᓗᖓᑕ ᓯᓚᑖᓂᑦ ᕿᕐᓂᖅᓯᖅᑲᑦᑕᖅᑐᑦ ᐱᕈᓅᕌᖓᒪᑕ. ᑲᓪᓚᖁᑎᓪᓕ (ᐅᖅᑲᐅᔾᔨᖏᑦ) ᐊᐅᐸᓪᓚᓐᓇᑦᑐᑰᓗᓪᓕᖅᐸᑦᑐᑎᒃ.

Kallait kallaqutillu growing in a tangle with grasses, cranberries, and *qijuktaaqpait* (Labrador tea). In the early fall, the variety of *kallait* (bear berry fruit) that grows around Aalasi's house turns black as it ripens. The *kallaqutit* (bearberry leaves) turn a blazing red.

ᑕᖅᑲᓚᐊᓘᔮᖅᑐᑎᒃ. ᐊᐅᐸᓪᓚᓇᒃᓯᖃᑦᑕᖅᑐᑦ ᐅᑭᐊᖅᓵᒃᑯᑦ. ᐱᕈᖅᓯᐊᕐᒋᑦ ᖁᖅᓱᕈᔪᒃᑐᑎᑦ ᓯᕗᓂᕈᑎᙳᐊᑯᓘᔮᖅᑲᑦᑕᖅᑐᑦ. ᑲᓪᓚᐃᓪᓗ ᐊᒻᒪᓗᑭᓵᓪᓚᓇᐅᕙᑦᑐᑦ. ᕿᑭᖅᑕᓂ ᓇᓂᔭᐅᓲᑦ ᕿᓂᖅᑑᕙᑦᑐᑦ ᐱᕈᓈᖅᓯᒪᓕᖅᑐᑎᒃ, ᑭᕙᓪᓕᕐᒥ ᓇᓂᔭᐅᖃᑦᑕᖅᑐᑦ ᐊᐅᐸᖅᓯᖕᒫᖃᑦᑕᖅᑐᑦ ᐱᕈᓈᖅᓯᒪᓕᖅᑎᓪᓗᒋᑦ. ᑕᒪᕐᒦᓪᓕ ᓇᓂᔭᐅᓲᖑᔪᑦ ᕿᑎᕐᒥᐅᓂ.

Kallait kallaqutillu is common in the southern Qikiqtani, some parts of the Kivalliq, and in many parts of the Kitikmeot. Kallait kallaqutillu can be distinguished from *kigutangirnait naqutillu* (blueberry) by the slightly jagged edges of its leaves (kigutangirnait naqutillu has smooth-edged leaves). The leaves are small with easily-visible veins. They turn bright red in the fall. The flowers are pale yellow-green and shaped like tiny bells. The kallait are round and smooth. The ones that are commonly found in the southern Qikiqtani are usually black when ripe, while the ones found in some parts of the Kivalliq are red when ripe. Both types are found in the Kitikmeot.

ᐊᐅᔭᐅᑉ ᕿᑎᖅᐸᓯᐊᓂ ᐃᖃᓗᐃᑦ ᓯᓚᑎᑦᑎᐊᕐᒪᓂ, ᑲᓪᓚᐃᑦ (ᐸᐅᖕᒪᖕᒋᑦ) ᐊᐅᐸᖅᑐᐃᑦ ᐱᕈᓈᓚᐅᖅᑎᓐᓇᒋᑦ ᐊᒻᒪᓗ ᑲᓪᓚᖁᑎᑦ (ᐅᖃᐅᔭᖕᒋᑦ) ᑐᖑᑦᑐᓇᑦᑐᑎᒃ.

Mid-season near Iqaluit, the *kallait* (bearberry fruit) are red before they ripen and the *kallaqutit* (bearberry leaves) are bright green.

ᑭᒎᑕᖏᕐᓇᐃᑦ ᓇᖁᑎᓪᓗ

Kigutangirnait Naqutillu / Blueberry

ᐃᓄᒃᑎᑐᑦ,ᑭᒎᑕᖏᕐᓇᐃᑦ (ᐊᑕᐅᓯᐅᓪᓗᓂ:ᑭᒎᑕᖏᕐᓇᖅ)ᐊᒻᒪ ᓇᖁᑎᑦ ᐅᖃᐅᓯᖃᖅᑐᑦ ᐅᖃᐅᓰᖏᓐᓂᑦ. ᑖᕝᕙᓂ ᑕᐃᒃᑕᓐᓂᐊᖅᑕᕗᑦ ᑭᒎᑕᖏᕐᓇᑦ ᓇᖁᑎᓗ.”). ᑭᒎᑕᖏᕐᓇᐃᑦ, ᑐᖁᖅᑐᖅ “ᑭᒎᑕᐃᕈᑏᑦ” ᐊᑎᖃᐅᑎᒋᒻᒪᒋᑦ ᑭᒍᑎᐅᑉ ᓱᕋᓂᖏᑦ ᑐᖑᔪᖅᑑᓕᖃᑦᑕᕐᒪᑕ ᐲᓚᐅᙱᓯᖕᒐᖅᑎᓪᓗᒋᑦ.

In Inuktitut, *kigutangirnait* (singular: *kigutangirnaq*) refers to the berries of the plant and *naqutit* refers to the leaves of the plant. Here, we refer to the plant generally as *kigutangirnat naqutillu* ("kigutangirnat and naqutit"). Kigutangirnait, which means "that which causes the teeth to be removed," are named for what it looks like when dark blue pieces remain on the teeth.

ᑭᒎᑕᖏᕐᓇᐃᑦ ᓇᖁᑎᓪᓗ ᓄᓂᕗᑦᑕᐅᔪᒪᓂᖅᐹᖑᕗᓪᓚᐃᔪᑦ. ᓄᓇᕙᒥᑦ ᐱᑕᖃᒐᔪᒪᑕ. ᑖᒃᑯᐊ ᐊᔾᔨᖏᑦᓚᙳᒪᒋᑦ ᐸᐅᖕᒐᐃᑦ ᐸᐅᖕᒐᖁᑎᓪᓗ ᐅᖃᐅᓯᖏᑕ ᑭᓪᓕᖏᑦ

ᑖᒃᑯᐊ ᑭᒎᑕᖏᕐᓇᐃᑦ ᐸᕝᓚᒃᐅᑎᓪᓗᒃᑐᖅ! ᐅᑭᐊᒃᓵᙳᓕᖅᒪᑦ ᓇᖁᑎᖏᑦ ᐊᐅᐸᕈᔫᒃᓯᔪᑦ, ᑖᕝᕙ ᐸᕝᓚᕐᑎᐊᓕᖅᑐᑦ ᑭᒎᑕᖏᕐᓇᐃᑦ.

These *kigutangirnait* are at their sweetest! Fall has arrived and the *naqutit* have turned into a soft red, which means the kigutangirnait are ripe.

ᑐᑭᓕᐊᑦᑎᐊᕐᒪᑕ (ᐸᐅᕐᖓᒐᐃᑦ ᐸᐅᕐᖓᕐᑯᑎᓪᓗ ᑲᐱᓪᓛᓇᖅᑐᑯᓘᒻᒪᑕ). ᐊᐅᔾᔭᒃᑯᑦ ᐅᖄᐅᔾᔪᑎᑦ ᑐᖑᑦᑐᕆᑦᑐᑎᒃ ᐊᐅᐸᓪᓚᕆᓕᖅᑲᑦᑕᖅᑐᑦ ᐅᑭᐊᒃᓴᑖᖓ. ᓇᑲᖏᑦ ᑕᑭᓪᓕᒻᒪᕆᖅᑲᑦᑕᖅᑐᑦ, ᑭᓯᐊᓂ 15 ᓴᐊᓐᑎᒦᑐᒥᑦ ᑕᑭᓂᖅᓴᐅᓕᕈᓐᓇᕐᑎᑦ. ᐱᕈᖅᓯᐊᖏᓪᓕ ᖃᑯᖅᑕᐅᖄᑦᑕᖅᑐᑦ ᐊᐅᐸᕈᔪᒃᑐᑎᓪᓗ, ᓯᕙᓂᕈᑏᕐᓛᙱᐊᑯᓘᔮᑐᑎᒃ. ᑭᒍᑕᖏᕐᓇᐃᑦ ᑐᖑᔫᖅᑑᖄᑦᑕᖅᑐᑦ ᐱᕈᕇᖅᓯᒪᓕᖅᑐᑎᒃ. ᐱᕈᓕᕆᖅᑎᓪᓗᒋᓪᓕ ᖃᑯᖅᑕᐸᓘᕙᑦᑐᑎᒃ.

Kigutangirnait naqutillu is perhaps the most sought-after berry plant. It is common in most parts of Nunavut. This plant is easily distinguishable from *kallait kallaqutillu* (bearberry) by the smooth edges of its leaves (kallait kallaqutillu has slightly jagged edges). The leaves are bluish-green in summer and turn bright red in autumn. The stems can grow quite long, but are usually less than 15 cm tall. The flowers are pink and white, shaped like tiny bells. Kigutangirnait are round and blue when ripe. Before they are ripe, they are almost white.

ᑭᒍᑕᖏᕐᓇᓂᑦ ᓄᓂᕚᓯᒃᑐᓂ ᐃᖏᐊᑦᑕᐅᖏᑦᑎᐊᕐᓗᓂ ᑭᓯᐊᓂᐅᕗᖅ ᐊᒻᒪ ᖀᓂᑦᑎᐊᒃᑲᐅᔨᐊᖅᑲᖅᑐᓂ. ᐅᖄᐅᔾᔪᓐᓄᑦ ᑕᖅᓴᐅᖏᑦᑐᑯᓘᖄᑦᑕᕐᒪᑕ ᐊᑖᓃᒍᑦᓗᑎᓪᓗ.

Collecting *kigutangirnait* requires dedication and a keen eye. The tiny treats are often hidden beneath their leaves.

ᑭᒍᑕᖏᕐᓇᕐᑯᑎᑦ ᓄᖑᕈᑎᖏᑦ ᕿᓚᒥᑯᓗᒃ
ᐊᐅᔭᒃᑯᑦ ᓴᖅᑮᔭᑲᐃᓐᓇᓲᑦ. ᖃᑯᖅᑕᐅᔪᑦ
ᐊᐅᐸᕈᔪᓐᓂᐊᕐᔪᖅᑲᖅᑐᑎᒃ.

The tiny blossoms of *kigutangirnait naqutillu* (blueberry) appear very briefly at the beginning of summer. They are white with delicate blushes of pink.

ᓇᕐᑯᑎᑦ (ᐅᕐᑲᐅᔭᖏᑦ) ᑭᓪᓚᖏᑦ ᒪᓂᕋᑦᑐᑦ,
ᑲᓪᓚᕐᑯᑎᓪᓕ (ᐅᕐᑲᐅᔭᖏᑦ) ᑭᓪᓚᖏᑦ
ᒪᓂᕋᓐᓇᑎᒃ. ᑖᓐᓇ ᐅᕐᑲᐅᔭᖅ ᓴᐅᒥᓐᓃᑦᑐᑦ
ᑎᑕᖅᑯᖓᓂᑦ ᑕᑯᔅᓴᐅᔪᖅ ᑲᓪᓚᕐᑯᑎᐅᔪᖅ
(ᐅᕐᑲᐅᔭᖏᑦ), ᐊᓯᖏᓪᓕ ᐅᕐᑲᐅᔭᐃᑦ ᑕᑯᔅᓴᐅᔪᑦ
ᓇᕐᑯᑎᐅᔪᑦ (ᑭᒍᑕᖏᕐᓇᐃᑦ ᐅᕐᑲᐅᔭᖏᑦ).

Naqutit (blueberry leaves) have smooth edges, while *kallaqutit* (bearberry leaves) have bumpy edges. The leaf at the bottom left-hand corner of this picture is kallaqutit (bearberry leaf), while the other leaves in this picture are naqutit (blueberry leaves).

ᐸᐅᕐᖓᓂᑦ ᐊᑐᕐᓂᐊᕐᓗᓂ
Using Berry Plants

ᐸᐅᕐᖓᐃᑦ ᐊᔾᔨᒌᖏᑦᑐᑦ ᐳᑯᑐᐃᓐᓇᕿᒃᓴᐃᓐᓇᐃᑦ. ᑭᒍᑕᖏᕐᓇᐃᑦ ᓯᕐᓇᓛᖑᓪᓗᑎᒃ, ᑭᓯᐊᓂ ᐸᐅᕐᖓᐃᑦ ᐊᒻᒪ ᑲᓪᓚᐃᑦ ᒪᒪᖅᑐᑯᓗᐃᑦ ᓂᕿᑦᑎᐊᕙᐅᓪᓗᑎᓪᓗ. ᐸᐅᕐᖓᐃᑦ ᑕᒪᒃᑯᐊᑦ ᖃᖓᑐᐃᓐᓇᒃᑯᑦ ᓂᕆᔭᒃᓴᐅᓲᖑᕗᑦ-ᐱᕈᓕᓵᖅᑎᓪᓗᒋᑦ ᐅᕝᕙᓘᓐᓃᑦ ᐱᕈᕇᖅᓯᒪᓕᖅᑎᓪᓗᒋᑦ. ᑭᒍᑕᖏᕐᓇᐃᑦ ᐸᐅᕐᖓᐃᓪᓗ ᐱᕈᕇᖅᓯᒪᓕᖅᑎᓪᓗᒋᑦ ᒪᒪᓛᙳᕗᑦ. (ᐅᑭᐊᒃᓴᓕᕐᑖᖅᑎᓪᓗᒍ), ᑭᓯᐊᓂᓕ ᐱᕈᕇᓚᐅᖅᑎᓐᓇᒋᑦ ᑲᓪᓚᐃᑦ ᒪᒪᕐᓂᖅᐹᖑᕐᑲᑦᑕᖅᐳᑦ.

The berries of all three plants can be eaten fresh from the plant. *Kigutangirnait* (blueberries) are the sweetest, but *paurngait* (crowberries) and *kallait* (bearberries) are also enjoyable and nutritious. All three types of berries can be eaten at any time—ripe or unripe. Kigutangirnait and paurngait taste best when they are ripe (in the early fall), but kallait are most delicious before they are ripe.

ᐱᓂᖅᓴᐅᕗᖅ ᓂᕆᓗᐊᑲᐅᑎᒋᖏᒃᑯᕕᑦ ᐸᐅᕐᖓᓂᑦ. ᐸᐅᕐᖓᐃᑦ ᐱᓗᐊᕐᓗᒋᑦ ᓈᙳᓕᕐᓇᕐᒪᑕ. ᓂᕆᔭᒃᓴᑑᔭᕆᐊᖃᖏᒃᑭᕗᑦ ᑳᓗᐊᖅᑐᒧᑦ ᓂᕆᑦᑎᐊᖃᑦᑕᖅᓯᒪᖏᑦᑐᒧᑦ. ᑭᓯᐊᓂᓕᑦᑕᐅ, ᑭᒍᑕᖏᕐᓇᕐᓂᖅ ᐃᓱᒪᐃᓐᓇᖅ ᓂᕆᕐᑯᔪᖅ.

It is best to avoid eating too many paurngait all at once. Eating too many paurngait may cause a stomach ache. Also, paurngait should not be relied on when a person is very hungry or malnourished. However, it is safe to eat as many kigutangirnait and kallait as you wish.

ᐸᐅᕐᖓᐃᑦ ᑭᒍᑕᖏᕐᓇᐃᓪᓗ ᓂᕆᓂᐊᖅᑕᓐᓄᑦ ᐃᓚᓕᐅᑎᔭᓐᓇᖅᑎᐊᖅᑕᐃᑦ. ᐅᖃᓕᒫᒐᕐᒥᑦ ᐊᑯᓕᐅᕈᑎᑦ ᒪᓕᓪᓗᒋᑦ ᓇᓕᐊᓐᓂᑐᐃᓐᓇᖅ ᐃᓚᓕᐅᑎᔭᓐᓇᖅᑖᒃᑭᑦ. ᓂᕿᓕᐅᕈᑎᓂᑦ ᐅᖃᓕᒫᒐᕐᒥᑦ ᐱᑕᖃᕐᒥᒻᒪᑦ ᐃᖃᓗᓕᐅᕈᑎᑦ, ᐸᐅᕐᖓᖃᖅᑐᒥᑦ, ᐊᒻᒪᓗ ᑲᑎᖅᓱᖅᓯᒪᔪᑦ ᐱᕈᖅᑐᕕᓂᑦ ᑐᖅᑕᓕᒃ ᑎᑎᖅᑯᖓ 63 (ᑭᒍᑕᖏᕐᓇᕐᓂᑦ ᑕᐅᖅᓱᖅᑕᐅᔪᓐᓇᖅᑐᑦ ᐸᐅᕐᖓᐃᑦ ᓂᕿᓕᐅᕈᑎᑦ ᒪᓕᓪᓗᒋᑦ). ᑭᒍᑕᖏᕐᓇᐃᑦ ᐸᐅᕐᖓᐃᓪᓗ ᐸᓚᐅᒑᓕᐊᒧᑦ ᐃᓚᓕᐅᔾᔭᐅᔪᓐᓇᖅᑐᑦ. ᑭᓯᐊᓂ ᐸᐅᕐᖓᓂᑦ ᐸᓚᐅᒑᓕᐊᑦ ᐃᓚᓪᓗᒋᑦ ᐱᐅᓂᖅᓴᐃᑦ ᑭᒍᑕᖏᕐᓇᐃᑦ ᐅᐊᑦᑎᐊᕈᓐᓇᓵᕈᒥᓇᕐᒪᑕ. ᐅᓪᓗᒥᐅᓕᖅᑐᖅᑐᓕ ᐸᐅᕐᖓᐃᑦ ᑭᒍᑕᖏᕐᓇᐃᓪᓗ ᓴᓪᓕᐊᖑᕙᓕᖅᑐᑦ ᐊᒻᒪ ᓯᕐᓇᖅᑐᓕᐊᖑᕙᓕᖅᑐᑎᑦ.

Paurngait and kigutangirnait can be made into many different dishes, such as *aluit* (puddings). There is also a recipe for a salad of char, paurngait, and *tuqtait* (alpine bistort rhizomes; p. 63). Kigutangirnait and paurngait can be added to bannock. But, Aalasi suggests that it may be best to use paurngait in bannock because fresh kigutangirnait are such a precious treat. These days, people also use paurngait and kigutangirnait in jams, jellies, and desserts.

ᐋᓚᓯ ᐃᖅᑲᐅᒪᔫᖅ ᐃᓚᒌᓚᒫᑦᑎᐊᑦ ᐸᐅᖕᒐᓂᑦ ᑭᒍᑕᖕᒋᕐᓇᓂᓪᓗ ᓄᓂᕗᖅᑲᑦᑕᓚᐅᖅᓯᒪᒐᒥ ᐊᐅᔭᐅᓕᖀᖕᒐᑦ. ᐊᒡᒐᓐᓄᑦ ᓄᓂᕗᒐᔅᓴᐅᔪᑦ, ᐅᓪᓗᒥᐅᓕᖅᑐᖅᑕ ᑭᓯᐊᓂ ᓄᓂᕗᐅᑎᖅᑲᖅᑲᑦᑕᓕᕐᒥᔪᑦ. ᓄᓂᕗᐅᑎᐅᑉ ᐊᑯᓐᓂᖕᒋᑎᒍ ᐸᐅᖕᒐᑦ ᐲᑦᑎᐊᖅᑲᑦᑕᖅᑐᑦ ᓇᑲᖕᒋᓐᓂᑦ.

Aalasi recalls that her entire family would work together to collect kigutangirnait and paurngait in the summer. The berries can be hand-picked, but some people prefer to use a rake-like tool called a *nunivauti*. The fingers of a nunivauti pull the berries off the stems all at once.

ᑲᓪᓚᖅᑯᑏᑦ, ᑲᓪᓚᐃᑦ ᐱᕈᖅᕙᖕᒋᑦ, ᒪᒪᖅᑐᒥᒃ, ᑏᓕᐊᕆᔭᒃᓴᐅᕗᑦ (ᐸᐅᖕᒐᐃᑦ ᐊᓱᖕᒋᑦ ᐅᖅᑯᔭᖕᒋᑦ ᑏᑦᑎᐊᕙᓕᐊᕆᔭᒃᓴᐅᖕᒋᓪᒪᑕ). ᐅᖅᑯᔭᖕᒋᑦ ᓇᑲᖕᒋᓪᓗ ᖃᖕᒍᐃᓐᓇᒃᑯᑦ ᓄᐊᑕᒃᓴᐅᕗᑦ, ᑐᖑᔪᖅᑕᐅᑎᓪᓗᒋᑦ ᑲᔫᑎᓪᓗᒋᓘᓐᓃᑦ. ᑲᓪᓚᖅᑯᑎᓂᑦ ᑏᓕᐅᕐᓂᐊᕈᕕᑦ, ᑎᒍᒥᐊᖀᒃᓴᓪᓗᐊᓂᑦ ᓇᑲᓂᑦ ᐅᖅᑯᔭᖅᑲᖅᓗᒋᑦ ᑎᖅᑎᑎᑦᑎᓗᑎᑦ ᐊᑯᓂᐊᓗᒃ. ᑲᓪᓚᖅᑯᑎᓂᑦ ᑏᓕᐊᓂᓯᒪᔪᑦ ᐋᓚᓯᐅᑉ ᒪᒪᓇᑖᑦᑎᐊᓇᕗᐃ.

Kallaqutiit, the leaves of bearberry, make a tangy, delicious tea (the leaves of the other two berry plants are not good for tea). The leaves and stems can be collected at any time, green or brown. To make kallaqutiit tea, boil a handful of stems with leaves for long time. Kallaqutiit tea is Aalasi's favourite tea.

ᕿᔪᒃᑖᖅᐸᐃᑦ
Qijuktaaqpait / Labrador Tea

ᐊᐅᔭᒃᑯᑦ ᕿᔪᑦᑖᖅᐸᐃᑦ ᐱᕈᓕᕌᖅᑎᓪᓗᒋᑦ ᖃᒃᑰᓪᓗᑎᒃ ᐅᓪᓗᓇᐊᑯᓘᔭᖅᑲᑦᑕᖅᑐᑦ ᓄᓇᔭᒥᑦ. ᓇᓗᓇᖏᑦᑐᑯᓘᓯᑦ ᑕᓪᓕᒪᓂᓪᓗ ᐱᕈᖅᑐᖅᑲᖅᑐᑎᒃ. ᐅᖅᑲᐅᔭᖏᑦ ᑐᖑᑦᑐᓇᑦᑐᑎᒃ ᐊᒥᑦᑐᑳᖑᔪᑦ ᐊᑖᓄᓇᐊᖓᖅᑰᔾᓪᓗᑎᒃ, ᒥᖅᑯᑎᙳᐊᑯᓘᔭᖅᑐᑦ. ᐊᑎᖏᑦ ᑲᔪᕈᔪᐃᑦ ᐊᐅᐸᕈᔪᑦᑐᑎᒃ ᒥᖅᑯᖅᑲᖅᑐᑎᓪᓗ. ᐱᕈᖅᕕᖏᑦ 15 ᓯᐊᓐᑎᒦᑐᓂᒃ ᑕᑭᓂᖅᑲᖅᑲᑦᑕᖅᑐᑦ.

In early summer, the blossoms of *qijuktaaqpait* look like bright white stars on the tundra. They have five white petals and grow in distinct clusters. The dark green leaves are narrow and the edges roll under, making them look like needles. Their undersides are covered with rust-coloured hairs. The stems grow in mats and are usually no longer than 15 cm.

ᕿᔪᒃᑖᖅᐸᐃᑦ ᑏᓕᐊᖑᒡᒍᔪᑦ ᐊᑲᐅᓯᔅᓴᐅᑎᖅᑲᓪᓚᓇᑦᑐᑦ. ᓈᖅᑐᑦᑐᓄᑦ ᐱᐅᓯᔾᔪᑕᐅᔪᓐᓇᖅᑐᑎᒃ ᐊᒻᒪᓗ ᖃᓂᒻᒪᑦᑕᐃᓕᒪᔾᔪᑕᐅᔪᓐᓇᒻᒪᑦᑐᑦ. ᖃᓂᒪᕈᔪᓚᐅᖅᑐᓂ ᑎᒥᒧᑦ ᒪᑭᑦᑎᐊᕈᑕᐅᔪᓐᓇᖅᑐᑦ ᐊᒻᒪᓗ ᑕᖅᑲᓴᕋᐃᖏᓐᓂᖅᓴᐅᓕᖅᑐᓂ.

Qijuktaaqpait is primarily used to make a powerful medicinal tea. It can help with severe stomach ailments and is said to have anti-microbial properties that can fend off infection. It is also known to be rejuvenating and to stimulate energy in those recovering from illness.

ᐱᔾᔪᑎᒋᓪᓗᒍ ᑭᓯᐊᓂ, ᓴᙱᓂᐊᓗᐊ, ᕿᔪᑦᑖᖅᐸᐃᑦ ᐊᔾᔨᒌᖕᒋᑦᑐᑦ ᐃᒻᒥᒃᑯᑦ ᑏᓕᐊᖑᖃᑦᑕᖕᒋᑦᑐᑦ ᐊᑲᐅᓯᓴᐅᑕᐅᓂᐊᕋᓗᐊᕈᓂ. ᑏᓕᐊᓇᓂᐊᕈᕕᐅᓪᓕ ᐃᓚᓕᐅᔾᔨᖕᒑᕋᔭᖅᑐᑎᑦ ᕿᔪᑦᑖᖅᐸᒥᑦ ᑏᓕᐊᓇᔭᓈᖅᓯᒪᔪᒧᑦ ᐸᐅᓐᓇᒻᒥᑦ, ᐊᑕᐅᓯᑯᑖᒥᒃ ᑏᓕᐊᓇᔭᐃᑦ ᐃᓗᐃᓐᓇᕐᓗᒍ (ᐸᐅᓐᓇᐃᑦ; p. 44) ᐊᓯᖕᒋᓐᓄᑦ ᓅᓐᓃᑦ ᐅᕝᕙᓘᓐᓃᑦ ᓂᐅᕕᖅᕕᒻᒦᙵᖅᑐᒥᑦ ᑏᒥᑦ. ᑏᓕᐊᓐᓄᑦ ᐴᑐᐃᓐᓇᕈᓐᓇᖅᑕᐃᑦ ᐅᕝᕙᓘᓐᓂᑦ ᑯᕕᔭᓈᖅᓯᒪᔭᓐᓄᑦ ᐃᖕᒍᓯᕐᒧᑦ ᑏᒧᑦ.

However, because of its potency, the Arctic variety of qijuktaaqpait should not be used to make medicinal tea on its own. To make a medicinal tea from qijuktaaqpait, add a single sprig of qijuktaaqpait to any other tea, such as *paunnait* (dwarf fireweed; p. 44) and other land teas or even store-bought teas. The sprig can be added for a few minutes to the teapot or right into the cup.

ᕿᔪᒃᑖᖅᐸᐃᑦ ᑏᓕᐊᓇᓂᐊᕐᓗᒍ ᒪᒪᕐᓂᖅᐹᖑᔪᖅ ᓴᙱᓂᖅᐹᖑᓪᓗᑎᓪᓗ ᑎᒥᒥᑦ ᐊᑲᐅᓯᓴᐅᑎᑦ ᓄᐊᑕᐅᒍᑎᒃ ᐅᑭᐊᒃᓵᒥᑦ. ᖁᐊᖅᑎᓯᒪᓯᖕᒐᕐᓗᒍ ᐊᕐᕌᒍᓚᒫᕐᒧᑦ ᐃᓚᒃᑯᐊᓇᔭᒃᓴᐅᔪᑦ ᓴᙱᓂᖃᑦᑎᐊᖕᒋᓐᓇᕐᓂᐊᕐᒪᑦ.

ᐊᐅᔭᓕᓵᒃᑯᑦ ᕿᔪᒃᑖᐸᐃᑦ ᐅᓪᓗᓇᐊᑯᓘᔮᖅᑐᖅ.

The starry blossoms of *qijuktaaqpait* (Labrador tea) in early summer.

Qijuktaaqpait makes the most delicious and medicinally powerful tea if it is collected in the fall. It can be used all year long if it is kept frozen to preserve its strength.

ᕿᔪᒃᑖᖅᐸᐃᑦ ᑭᓪᓚᒧᑦ ᓴᓗᒪᖅᓴᐃᔾᔪᑕᐅᔪᓐᓇᕆᕗᑦ. ᓴᓗᒪᖅᓴᐃᔾᔪᑎᒋᓂᐊᕐᓗᒍ, ᑎᖅᑎᑦᑐᒦᑎᑲᐃᓐᓇᕐᓗᒋᑦ ᑎᒍᓪᓗᐊᕋᒃᓴᐃᑦ. ᐃᒪᖅ ᑎᓴᐃᕐᔫᒥᑉᐸᑦ ᖃᓪᓗᓈᖅᑕᒥᑦ ᒪᓯᑦᑎᓇᓗᑎᑦ ᖃᔪᖕᒐᓂ ᑖᓐᓇ ᑭᓪᓕᕐᒥᑦ ᓴᓗᒪᖅᓴᐃᔾᔪᑎᒋᓕᕐᓗᒍ. ᕿᔪᒃᑖᐸᐃᑦ ᐊᑲᐅᓯᓴᐅᑎᖏᑦ ᑭᓪᓕᕐᒥᑦ ᓱᕈᙱᑎᑦᑎᔪᓐᓇᕐᒪᑦ.

Qijuktaaqpait can also be used to clean wounds. To use it as a cleanser, boil a handful in a pot of water. When the water has cooled enough, soak a cloth in it and use the cloth to clean the wound. The anti-microbial properties of qijuktaaqpait will decrease the risk of infection.

ᐊᒥᓱᓄᑦ ᕿᔪᒃᑖᐸᐃᑦ ᑎᐱᖕᒐ ᒪᒪᕆᔭᐅᒻᒪᑦ. ᐃᑦᓗᓐᓄᑦ ᓇᒧᑐᐃᓐᓇᖅ ᐃᓕᐅᖅᑲᕈᓐᓇᕐᒥᔭᐃᑦ ᑎᐱᑦᑎᐊᓇᒍᑕᐅᓂᐊᕐᒪᑦ.ᑎᐱᖅᑲᑦᑎᐊᕈᓐᓇᕐᓂᖅᐹᖑᔪᖅ ᐃᒻᒫᕐᔪᔅᓯᒪᓗᒋᑦ ᐸᓂᖅᑎᑦᑕᐃᓕᒪᓗᒋᑦ.

Many people find the spicy scent of qijuktaaqpait to be soothing. It can be placed throughout the home to be enjoyed. It will smell best if it is set in a small amount of water to keep it moist.

ᐊᐅᔭᐅᔅᓴᓕᖅᑎᑦᓗᒍ ᕿᔪᒃᑖᐸᐃᑦ. ᐅᑦᓗᓇᖑᔮᓚᐅᖅᑐᖅ ᐱᑕᖅᑲᕈᓐᓃᖅᓗᐊᖅᑎᓗᒋᑦ ᑭᓯᐊᓂ ᐅᖅᑲᐅᔾᔨᖕᒋᓐᓄᑦ ᓯᓚ ᓇᓂᔪᓐᓇᖅᔭᖅᑕᐃᑦ.

Late summer *qijuktaaqpait*. The starry white blossoms have gone but you can still find this plant by its distinctive leaves.

ᕿᔪᒃᑖᕐᐸᐃᑦ ᓴᙱᓂᕐᐹᖑᔪᑦ ᒪᒪᕐᓂᕐᐹᖑᓪᓗᑎᓪᓗ ᐅᑭᐊᒃᓵᒃᑯᑦ.

Qijuktaaqpait is most delicious and medicinally powerful in the early fall.

ᓯᐅᕋᐅᑉ ᐅᖃᐅᔭᖏᑦ
Siuraup Uqaujangit / Seaside Bluebells

ᓯᐅᕋᐅᑉ ᐅᖃᐅᔭᖏᑦ ᓯᔾᔮᒥᓲᑦ. ᐅᖃᐅᔭᖏᑦ ᐊᓅᑎᑎᑐᑦ ᒥᒃᓯᖅᑲᖅᑐᑎᑦ ᓯᐊᕐᓇᕈᔫᓪᓗᑎᒃ ᑐᖑᔪᖅᑕᐅᓪᓗᑎᒃ ᐊᒻᒪᓗ ᕿᓪᓚᕋᑎ. ᓯᓚᓐᓂᖏᑦ 3-5 ᒥᓚᒦᑐᒦᖅᑲᑦᑕᖅᑐᑦ. ᐱᕈᖅᓯᐊᕚᔾᔪᑯᓗᖏᑦ ᑐᖑᔪᕈᔪᑦᑐᑯᓘᖅᑲᑦᑕᖅᑐᑦ,ᓯᕙᓂᕈᑎᑐᑦ ᒥᒃᓯᖅᑲᖅᑐᑎᒃ. ᓯᐅᕋᐅᑉ ᐅᖃᐅᔭᖏᑦ ᑕᑭᓪᓚᓗᐊᕈᓐᓇᖏᑦᑐᑦ, ᑭᓯᐊᓂᑦ ᓇᑲᖏᑦ 30 ᓯᐊᓐᑎᒦᑐᓂᑦ ᑕᑭᓂᖅᑲᓚᓲᑦ.

Siuraup uqaujangit grows in tufts on ocean beaches. The spoon-shaped leaves are light gray-green and they are matte (not shiny). They are usually 3 mm to 5 mm thick. The little flowers are light blue, shaped like tiny bells. Siuraup uqaujangit does not grow tall, but its stems can grow up to 30 cm horizontally.

ᓯᐅᕋᐅᑉ ᐅᖃᐅᔭᖏᑦ ᓴᙱᓗᐊᕋᑎᑦ ᒪᒪᖅᑐᑯᓗᐃᑦ. ᓲᕐᓇᕈᔪᑦᑐᑎᒃ ᐊᕿᑦᑐᑯᓘᓪᓗᑎᒃ ᐊᒻᒪ ᐃᒃᓯᖅᑲᑦᑎᐊᖅᑐᑎᒃ. ᐋᓚᓯᓚ ᐃᖅᑲᐅᒪᔪᖅ ᑕᐃᒪᐃᑦᑐᓂ ᓂᕇᓐᓇᕈᓐᓇᖅᑲᑦᑕᓚᐅᕐᓂᖓᓂᑦ ᓂᕕᐊᓯᐊᖑᑦᓗᓂ. ᐅᖃᐅᔭᖏᑦ, ᐱᕈᖅᓯᐊᖓ, ᐊᐃᕋᖏᓐᓄᑦ ᓂᕆᔭᒃᓴᐃᓐᓇᐅᕗᑦ. ᒪᒪᕐᓂᖅᐹᖑᓪᓚᐃᔪᖅ ᐳᑯᑐᐃᓐᓇᕆᐊᖓ, ᖃᓄᐃᓕᖏᑦᑎᐊᕐᓗᒍ.

Siuraup uqaujangit is mild and pleasant to eat. It has a slight sweetness and is very tender and juicy. Aalasi recalls that there was no limit to how much she could eat of this plant as a child. The leaves, flowers, stems, and roots are all edible. The most delicious way to eat this plant is freshly-picked, just as it is.

ᓯᐅᕋᐅᑉ ᐅᕐᑲᐅᔭᖏᑦ ᒪᒪᕐᑐᐊᓘᒻᒥᔪᑦ ᐱᕈᕐᑐᕕᓂᕐᓂᑦ ᑐᖑᔪᕐᑕᓂᑦ ᐃᓚᓕᐅᑎᓗᒋᑦ, ᓲᕐᓗ ᐸᐅᓐᓇᓄᑦ (p. 44) ᐅᕝᕙᓘᓐᓃᑦ ᕐᑯᖑᓕᕐᓄᑦ (p. 52). ᐅᕐᑲᐅᔭᖏᑦ ᓇᑲᖏᓪᓗ ᐃᓚᓕᐅᔾᔭᐅᔪᓐᓇᕐᒥᔪᑦ ᐃᕐᑲᓗᕕᓂᐅᑉ ᕐᑲᔪᖕᒪᓄᑦ. ᐊᒃᑐᒪᓖᓚᐅᕐᓗᒋᑦ ᕐᑲᔪᕐᒧᑦ ᐃᓚᓕᐅᑎᓗᒋᑦ ᐃᕐᑲᓗᕕᓂᕐᑕᕐᑲᕈᓐᓃᕐᑎᓪᓗᒍ. ᐊᐃᕋᖏᑦᑕᐅᑉ ᓂᕆᔭᐅᔪᓐᓇᕐᒥᔪᑦ ᐆᓯᒪᓗᒋᑦ, ᑐᕐᑕᑎᑐᑦ (p. 58). ᑎᕐᑎᑎᓚᐅᑲᓪᓗᒋᑦ ᕐᑭᓚᒥᕈᓗ ᐊᕐᑭᓪᓕᒐᔭᕐᑐᑦ ᓴᙱᓗᐊᕈᓐᓃᕐᓗᑎᓪᓗ.

Siuraup uqaujangit is also delicious mixed with other greens, such as *paunnait* (dwarf fireweed; p. 44) or *qunguliit* (mountain sorrel; p. 52). The leaves and stems can also be used to season and enrich fish broth. Chop them into small pieces and add them to the broth that remains after the boiled fish has been removed. The roots can also be eaten

lightly cooked, as with *tuqtait* (alpine bistort rhizomes; p. 58). Cook them briefly in boiling water and they will become mild and tender.

ᓯᐅᕋᐅᑉ ᐅᖃᐅᔭᖏᑦ ᐊᑲᐅᓯᓴᐅᑎᐅᕗᓐᓇᕐᒥᕗᑦ. ᐸᓐᓂᖅᑑᒥᑦ ᐊᕐᓇᕐᒥᑦ ᐊᐅᖏᕈᑎᕗᖅᑲᓚᐅᖅᓯᒪᒻᒪᑦ. ᐋᓚᓯᐅᑉ ᐊᓈᓇᖓᑕ ᓂᕆᑎᙵᖏᖕᒎᖅᑐᓂᐅᑉ ᓯᐅᕋᐅᑉ ᐅᖃᐅᔭᖓᓂ ᖃᕗᖅᑐᖅᑎᖅᑲᑦᑕᓚᐅᖅᓯᒪᕘ. ᐊᑯᓂᐅᖏᑦᑐᖅ, ᐊᕐᓇᖅ ᐊᑲᐅᓯᕗᕕᓂᖅ. ᑏᓐᖑᐊᓕᐅᕐᓂᐊᕐᓗᓂ, ᐅᖃᐅᔭᖏᑦ ᓇᑲᖏᓪᓗ ᖁᓚᒥᑯᓗᒃ ᑎᖅᑎᑦᑐᒦᑎᒋᐊᓚᑦ, ᓴᙱᕗᒻᒪᕆᐅᒻᒪᑕ.

Siuraup uqaujangit also has medicinal properties. There was once a woman in Pangnirtung who had lost a lot of blood. Aalasi's mother replaced the woman's meals with tea made from siuraup uqaujangit. After a short while, the woman recovered. To make tea from siuraup uqaujangit, boil the leaves and stems only briefly, as they are strong in flavour when boiled.

ᐱᕈᖅᓯᐊᕌᕐᕗᑯᓗᖏᑦ
ᓯᐅᕋᐅᑉ ᐅᖃᐅᔭᖏᑦ
ᖁᓚᒥᑯᓗᒃ
ᓴᖅᑭᔫᑲᐃᓐᓇᔫᑦ
ᐊᐅᔭᐅᑉ ᖁᑎᐊᒍ
ᐱᑕᖃᕈᓐᓃᓈᖃᑦᑕᖅᑐᑦ.

The tiny blossoms of *siuraup uqaujangit* appear briefly in mid-summer and then disappear.

ᐃᑦᑎᓇᖅᑐᖅ ᐃᖃᓗᕕᓂᐅᑉ ᖃᔪᖓ ᓯᐅᕋᐅᑉ ᐅᖃᐅᔾᔨᖏᓐᓂᑦ ᐃᓚᓕ.

Savory Char Broth with Siuraup Uqaujangit

- 1 ᐃᕐᖑᓯᖅ ᓯᐅᕋᐅᑉ ᐅᖃᐅᔾᔨᖏᑦ, ᐊᒃᑯᑲᓛᔅᓯᒪᔪᑦ
- ᐃᖃᓗᕕᓂᐅᑉ ᖃᔪᖓ

ᐃᖃᓗᓕᐅᕐᕕᕕᓃᑦ ᖃᔪᖓ (ᐃᑎᐅᔭᖅᑲᕈᓐᓇᖅᑐᖅ ᐊᒡᒐᓘᓐᓃᑦ). ᖃᔪᖅ ᐆᓇᕐᓂᖓᓂ, ᓯᐅᕋᐅᑉ ᐅᖃᐅᔾᔨᖏᑦ ᐊᒃᑯᑲᓛᒃᓯᒪᔪᑦ ᐃᓚᓕᐅᑎᓗᒋᑦ. ᐃᖑᓚᕐᓗᒍ ᐃᑲᕐᖑᐊᒥᑦ ᐊᑕᐅᓯᕐᒥᑦ ᐊᒻᒪ ᖃᔪᖅᑐᖅᑎᐊᓕᕐᒥᓗᑎᑦ!

ᖃᐅᔨᒪᔭᕆᐊᓕᒃ: ᓯᐅᕋᐅᑉ ᐅᖃᐅᔾᔨᖏᑦ ᓴᙱᔪᒻᒪᕆᐅᒻᒪᑕ ᖃᔪᖅᑐᕐᓇᓵᖅᑕᐃᓕᒍ ᕿᑲᑯᑖᑦᑎᑦᑕᐃᓕᐅᑉ ᓴᙱᓗᐊᓕᑐᐃᓐᓇᕆᐊᖃᕐᒪᑦ.

- 1 cup of siuraup uqaujangit, chopped into small pieces
- Pot of cooking water from boiled fish

Save the cooking water from a pot of boiled char (with or without onions). While the water is still hot, add a generous handful of the chopped leaves and stems of siuraup uqaujangit. Stir the mixture for a minute or so and enjoy!

Note: siuraup uqaujangit is potent, so do not wait too long to drink the broth or the flavour may become too strong.

ᒪᓕᒃᑳᑦ
Malikkaat / Mountain Avens

ᒪᓕᒃᑳᑦ ᓄᓇᕋᖅᑖᓰᖑᕗᑦ ᔫᓇ ᐊᒻᒪᓗ ᔪᓚᐃ ᓄᙳᐊᓂ. ᒪᓕᒃᑳᑦ ᖃᑯᖅᑕᐅᖃᑦᑕᖅᑐᑦ ᖁᖅᓯᖅᑐᓂᑦ ᕿᑎᖃᖅᑐᑎᑦ. ᐊᒻᒪᓗᖅᑐᑯᓘᔪᑦ ᑖᒃᑯᐊ. ᒪᓕᒃᑳᑦ ("ᒪᓕᑦᑏᑦ") ᓯᖅᑭᓂᕐᒧᑦ ᓵᙵᓰᑦ, ᐅᑦᑐᑕᓘᑦᑎᐊᖅ ᐃᖕᒋᕋᓂᖕᒐᓂᑦ ᒪᓕᖅᑲᑦᑕᖅᑐᑦ. ᓇᑲᖕᒋᑦ ᒥᖅᑯᕈᔪᖅᑲᖅᑐᑎᒃ ᐊᒻᒪ ᐅᖅᑲᐅᔾᔭᖅᑲᕋᑎᒃ, ᑕᕿᓂᖅᑲᒡᒍᔪᑦ 4 - 12 ᓯᐊᓐᑎᒦᑕᓂᑦ. ᓄᓇᐅᑉ ᐊᑕᔫᖅᓯᒪᔪᓂᒃ ᐋᖅᓂᑦ ᐱᕈᖅᑲᑦᑕᖅᑐᑦ ᓴᒃᑯᙳᐊᕋᑖᑯᓗᒃᔮᖅᑲᑦᑕᖅᑐᑦ.

Malikkaat flowers in late June or July. The flowers are creamy white with yellow centres. They are shaped like tiny satellite dishes. Malikkaat ("the one that follows") turn to face the sun, following it across the sky all day. The flowering stems are hairy and leafless. They are usually 4 cm to 12 cm high. The leaves grow on stems along the ground and are shaped like small arrow heads.

ᐱᕈᖅᓯᐊᖕᒋᑦ ᐱᕈᓚᐅᖅᑎᓪᓗᒋᑦ, ᒥᖅᑯᖅᑳᖅᑲᑦᑕᖅᑐᑦ ᓂᒪᐅᑎᓯᒪᔪᑯᓗᓐᓂᒃ ᕿᑎᐊᓂᙶᖅᑐᓂᒃ ᓇᑉᐸᖕᒐᔪᓂᒃ. ᐅᑭᐊᒃᓴᓕᖅᑎᓪᓗᒍ, ᐱᕈᖅᓯᐊᖕᒋᑦ ᑲᑕᒐᖅᑲᑦᑕᖅᑐᑦ ᐊᒻᒪᓗ ᒥᖅᑯᑯᓗᖕᒋᑦ ᐸᓂᖀᖕᒐᑕ ᓂᒪᐅᑎᓯᒪᔪᓐᓃᖅᑐᑎᒃ ᐅᖅᑯᖕᒐᓕᖅᑲᑦᑕᖅᑐᑦ, ᐊᓄᕆᖅᒧᓗ ᑎᑦᑕᐅᑎᑕᐅᕙᑦᑐᑎᒃ.

After the plant flowers, hairs from the centre of the flower grow up into twisted spikes. Toward autumn, the petals fall away, the seeds ripen, and the hairs uncoil, spreading the seeds in the wind.

ᒪᓕᒃᑳᑦ ᖃᐅᔨᒪᔾᔪᑎᐅᔪᓐᓇᖅᑐᑦ ᓯᓚᒥᑦ. ᒥᖅᑯᖏᑦ ᓂᒪᐅᑎᓯᒪᔪᓐᓃᓕᕋᖓᑕ, ᑕᐃᒪ ᐅᑭᐊᒃᓵᓕᖅᒥ ᐅᑭᐅᕐᒧᑦ ᐸᕐᓇᐸᑦᑕᐊᓇᖅᓯᑦᑐᓂ. ᒪᓕᒃᑳᑦ ᓯᓚᐅᑉ ᖃᓄᐃᓐᓂᕆᓂᐊᖅᑕᖓᓂ ᖃᐅᔨᒪᔾᔪᑎᐅᔪᓐᓇᕆᐹᑦ. ᐱᕈᖅᓯᐊᑯᓗᖏᑦ ᓱᒃᖑᖅᑰᔾᔭᑦᑐᑎᒃ ᐃᓵᖑᑦᑐᑎᒃ ᑲᑎᒻᒪᑦᑎᐊᕌᖓᑕ, ᑕᐃᒪ ᕿᓯᐊᓂ ᓂᑦᑕᓱᓐᓂᐊᕐᒥ. ᐱᕈᖅᓯᐊᖏᑦᑐ ᖃᓱᖑᑦᑐᑎᒃ ᒪᐱᖕᖓᑦᑎᐊᖅᑎᑦᑐᕆᑦ, ᑕᐃᒪᔫᖅ ᓯᓚ ᐅᖅᑰᓂᐊᕐᒪᑦ.

Malikkaat can be used to judge the season. When the hairs start to uncoil, it is almost autumn and time to prepare for winter. Malikkaat can also be used to judge the weather. If the blossoms are tightly cupped, then the weather will be cold. If the blossoms are open and loose, the weather will be hot.

ᐊᔾᔨᓕᐅᕆᔪᕕᓂᖅ ᐃᐊᓚᓐ ᓯᒡᓗᕐ Photo by Ellen Ziegler

ᒪᓕᒃᑳᑦ ᐱᕈᓕᓲᑦ ᔪᓚᐃᕈᕋᑖᖅᑎᓪᓗᒍ.

Malikkaat bud in early July.

ᒪᓕᒃᑳᑦ ᔪᓚᐃ ᕿᑎᐊᓂᑦ ᐱᕈᓅᖅᑐᖅ.

Malikkaat blossom in mid July.

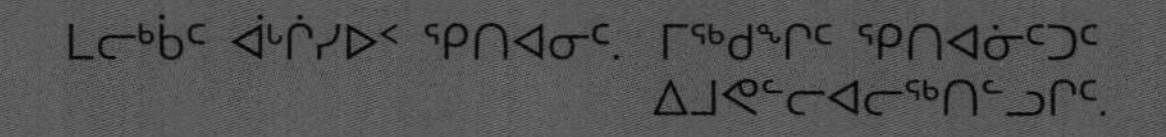

ᒪᓕᒃᑳᑦ ᐋᒡᒌᓯᐅᑉ ᕿᑎᐊᓂᑦ. ᒥᖅᑯᖏᑦ ᕿᑎᐊᓅᑦᑐᑦ
ᐃᒧᕙᑦᓕᐊᓕᖅᑎᓪᓗᒋᑦ.

Malikkaat blossom in mid-August. The hairs are beginning to coil up at the centre.

ᒪᓕᒃᑳᑦᑕᐅᖅ ᖃᐅᔨᒪᔾᔪᑕᐅᔪᓐᓇᕆᓗᑦ ᓇᓕᐊᓐᓄᑦ ᑐᕌᓇᐊᖅᑲᒪᖓ ᐊᖏᕐᕋᕐᓱᒃᑐᑦ ᐃᓄᒃ ᑕᒡᓯᓕᐅᔾᔭᐅᓯᒪᔪᖅ ᐅᕙᓂᓐᓂᑦ ᑖᖅᓯᔪᐊᓗᓐᓂᕈᓂ. ᐱᕈᖅᓯᐊᖑᓂᖓ ᓯᕿᓂᕐᒧᑦ ᓵᙵᐅᔭᐃᓐᓇᕐᒪᑦ, ᓇᒦᓰᖑᒻᒪᖔ ᖃᐅᔨᒪᒍᕕᐅ ᓯᕿᓂᖅ ᖃᖓᒃᑯᑦ, ᖃᐅᔨᔾᔪᑎᒋᔪᓐᓇᕋᔭᖅᑕᑎᑦ, ᓇᓕᐊᓐᓄᑦ ᓵᙵᓗᑎᑦ ᐱᓯᒋᐊᖅᑲᒪᖔᖅᐱᑦ.

Malikkaat can also help someone if they are lost in fog or if it is dark and they cannot see their way home. The buds and flowers always turn to face the sun, so if you are familiar with the sun's location in relation to your home, you will know which direction to take.

ᐃᑲᔪᕐᓂᖅᑲᕐᓂᖏᓐᓄᑦ ᒪᓕᒃᑳᑦ, ᐃᓄᐃᑦ ᐅᖃᐅᔾᔭᐅᖃᑦᑕᖅᑐᕗᓂᑦ ᐱᔮᓇᓗᑎᒃ ᑐᓕᕆᖃᑦᑕᕐᑯᔭᐅᓇᑎᒃ. ᐊᒻᒪᑦᑕᐅ, ᐊᕐᓇᐃᑦ ᐃᕐᓂᓕᕌᖓᑕ, ᐋᓚᓯ ᐃᖅᑲᐅᒪᒻᒥᔪᖅ ᖃᐅᔾᔩᖃᑦᑕᓚᐅᖅᓯᒪᒐᒥ ᐃᕐᓂᐊᖑᔫᑉ

ᒪᓕᒃᑳᑦ ᐃᒧᓯᒪᔪᓐᓂᖅᐸᑦᑕᓕᐊᔪᖅ ᓲᑎᐱᓇᐅᑉ ᐱᒋᐊᕐᓂᖓᓂᑦ.

Malikkaat beginning to uncoil in early September.

ᓂᐊᖏᓐᖓ ᒪᓕᒃᑳᓄᑦ ᐅᐃᔾᔭᓪᓚᒃᑎᑕᐅᖅᑲᑦᑕᕐᓂᒪᑕ ᐃᕐᓂᓱᒃᓲᔪᓄᑦ, ᓯᕿᓂᐅᑉ ᐃᖏᕐᕋᓂᖓᓄ (ᓯᕿᙵᒍᔭᐅᑉ ᐅᐃᔾᔮᕐᓂᖓ ᒪᓕᒃᑐᒍ).

Because malikkaat are so helpful, people are told not to step on them or destroy them on purpose. Aalasi recalls that when women had babies it was common for midwives to take the baby's head and gently turn it in the direction of malikkaat, toward the sun (clockwise).

ᓇᑲᖓᑕ ᒥᖅᑯᓯᔪᖏᑦ ᐊᑐᕐᓗᒋᑦ ᐃᔨᑎᓯᑎᒋᔭᐅᖅᑲᑦᑕᕐᓂᕆᕙᑦ. ᑕᐃᒫᓪᓚ ᐊᑐᖅᑕᐅᓂᐊᖅᑎᓪᓗᒋᑦ, ᒪᓕᒃᑳᑦ ᐃᖏᕐᕋᓂᕆᖅᑲᑦᑕᖅᑕᖓ ᒪᓕᒃᑐᒍ ᐅᐃᔾᔭᓪᓚᒃᑎᑕᐅᖅᑲᑦᑕᕐᓂᕆᕙᑦ.

The hairs on the stems of malikkaat may be used like tiny cloths to clean an irritation from an eye. When using the stems this way, they are to be turned in the direction that the malikkaat turn to follow the sun.

ᒪᓕᒃᑳᑦ ᐃᒧᓯᒪᖏᑦᑎᐊᓕᖅᑐᖅ ᐅᑭᐊᒃᓴᖅᑎᓪᓗᒍ, ᐊᐱᓂᐊᑦᑎᐊᓕᖅᑎᓪᓗᒍ.

Malikkaat fully uncoiled in autumn, just before the first snowfall.

ᐅᕐᔪ

Urju / Peat Moss

ᐅᕐᔪ ᐱᑕᕐᑲᒐᔪᒃᑐᑦ ᑕᓱᐅᔭᐃᑦ ᑭᑦᓕᖏᓐᓂᑦ ᐊᒻᒪᓗ ᒪᓴᕐᓗᒻᒥᑦ, ᐱᓗᐊᕐᑐᒥᑦ ᓄᓇᕗᑦ ᐅᕐᑯᕐᐸᓯᖕᒐᓂᑦ. ᐱᕈᕐᑲᑦᑕᕐᑐᑦ ᒪᓴᕈᔪᒻᒥᑦ, ᓄᓇᓂᑦ ᐊᕿᑦᑐᓂᑦ. ᐊᒥᐊᕐᑲᓲᖑᑦᓗᑎᑦ ᑲᔪᕐᓂᑦ, ᑐᖑᔪᕐᑕᓂᑦ, ᖁᕐᓱᕐᑐᓂᑦ ᐊᒻᒪᓗ ᖃᑯᕈᔪᒃᑐᓂᑦ.

Urju is common around the edges of ponds and in other wet areas of the tundra, especially in the southern parts of Nunavut. It grows in moist, spongy mats. Its colour ranges from brown and green to yellow and creamy white.

ᐅᕐᔪ ᓴᓗᒻᒪᕐᓴᐅᑎᕐᑲᕐᐳᑦ ᐊᒻᒪ, ᐸᓂᕐᑎᓯᒪᑦᓗᒥᑦ ᐃᒃᓯᑖᒃᑲᓐᓂᕈᓐᓇᕐᑐᑎᑦ ᐊᑦᑕᕈᑎᒥᓗᒥᑦ. ᑕᐃᒫ, ᐋᑑᑎᑕᓪᒪᓇᐅᒐᑦ.

Urju has natural antiseptic properties and, when it is dried, it is highly absorbent. So, it has many practical uses.

ᐋᑕᓯ ᐃᕐᑲᐅᒪᒍᑦ ᐃᑕᒦᑦ ᐊᖏᔪᒻᒪᓇᓐᓂᑦ ᓄᐊᑦᑎᕐᑲᑦᑕᓚᐅᕐᓯᒪᒐᒥᑦ ᐅᑭᐅᒃᑯᑦ ᐃᑕᒃᑯᐊᔅᓴᖏᓐᓂᒃ. ᐅᕐᔪ ᐸᓂᕐᓯᐊᓇᔭᐅᕐᑲᑦᑕᓚᐅᕐᓯᒪᔪᑦ ᐊᑐᓇᐊᕐᑲᑖᖕᒐᑕ ᐊᑐᐃᓐᓇᐅᓂᐊᕐᒪᑕ. ᐸᓂᕐᔫᓂᐊᕐᓗᓂ ᐅᕐᔪᓂᑦ, ᓄᐊᑦᑎᑕᐅᕐᓗᑎᑦ ᐅᔭᖁᐃᑦ ᖄᖕᒐᓄᑦ ᐅᖖᓘᓐᓃᑦ ᐸᓂᕐᑐᒧᑦ ᐃᑕᐅᕐᑐᕈᓐᓇᕐᑕᑎᑦ. ᑕᐃᒪᙵᑕᓖᐸᓗᒃ, ᐅᑦᓗᐃᓐᓇᕐ ᐸᓂᖁᓱᓲᑦ. ᐅᕐᔪ ᐸᓂᕐᑎᓯᒪᑦᓗᒥᑦ ᓂᕈᒥᑦᑎᒥᔪᑦ ᐅᑲᑕᐅᑉ ᒥᕐᑯᖕᒐᑎᑐᑦ.

Aalasi recalls her family collecting large quantities of urju in the summer to be used during the winter. The urju was dried and then stored in a place to which her family could return when they needed some. To dry urju, pick it from the ground and spread it across rocks or another dry surface. Normally, it will dry within a day. Dried urju is as soft as rabbit fur.

ᐸᓂᖅᑎᓯᒪᔪᑦ ᐅᕐᔪ ᖁᑦᑕᕐᕕᐅᔪᓐᓇᕐᒥᔪᑦ ᐅᕝᕙᓗᓐᓃᑦ ᐊᐅᓈᖅᑐᒧᑦ ᐊᑐᖅᑕᐅᔪᓐᓇᖅᑐᑎᒃ. ᖁᑦᑕᕐᕕᐅᓂᐊᖅᑎᑦᑐᒋᑦ ᐊᐅᓈᖅᑐᒧᑦᓘᓐᓃᑦ ᐊᑐᖅᑕᐅᓂᐊᖅᑎᑦᑐᒋᑦ, ᐋᓚᓯ ᐃᖅᑲᐅᒪᔪᖅ ᐸᓚᐅᒑᖅ ᐹᕕᓂᖕᒐᓂᑦ ᖄᖅᑲᖅᑎᑕᐅᖅᑲᑦᑕᓚᐅᕐᓂᖕᒋᓐᓂᒃ. ᐸᓚᐅᒑᖅ ᐹᕕᓂᖕᒋᑦ ᓂᕈᒥᒻᒪᑕ ᐅᐊᓴᕐᓂᖅᑐᑎᑦᑐ. ᖁᑦᑕᕐᕕᓕᐅᕐᓯᒃᑐᓂᓕ ᖃᑦᑐᓈᖅᑕᖅ ᓴᓇᖅᑳᓚᐅᖅᑐᒍ ᐃᓗᑦᓕᑐᐃᓐᓇᖅᑲᑦᑕᖅᑕᕕᓂᖕᒋᑦ ᐅᕐᔪᓂᒃ. ᑎᒍᓯᓗᑎᑦ ᒪᕈᐊᖅᑎᕐᓗᒍ ᐊᒡᒐᐃᑦ ᑕᑦᑕᓗᒍ ᐊᑕᐅᓯᕐᒧᑦ ᖁᑦᑕᕐᕕᒻᒧᑦ ᓈᒻᒪᖅᑲᑦᑕᖅᑐᕕᓂᖅ. ᐋᓚᓯ ᐃᖅᑲᐅᒪᔪᖅ ᐅᕐᔪ ᐊᑐᐃᓐᓇᐅᑎᓐᓇᒋᑦ ᐅᑲᓕᐅᑉ ᐊᒥᖕᒐᓂᑦ ᐊᑐᖅᑲᑦᑕᖅᑐᕕᓂᖅ, ᑮᓯᐊᓂ ᐅᕐᔪ

ᐋᓚᓯ ᖃᓄᕋᐅᑎᕋᓛᒧᑦ ᐲᖅᓯᔪᖅ ᐅᕐᔪᒥᑦ ᑕᓯᐅᑉ ᑭᑦᓕᖕᒐᓂᑦ.

Aalasi uses a small shovel to remove *urju* from the wet edges of a pond.

ᐱᐅᕆᓂᖅᓴᓇᖅᑲᑦᑕᖅᑕᕕᓂᖏᑦ ᐃᒋᐅᐃᓐᓇᕋᖅᓴᐅᓂᖏᓐᓄᑦ ᐊᑐᓈᖅᑐᕆᑦ. ᐊᓪᓚᓯ ᐃᖅᑲᐅᒪᔪᖅ ᐅᖅᔪᐃᑦ ᐊᓪᓕᓂᓕᐊᖑᖅᑲᑦᑕᓚᐅᖅᓂᖏᓐᓂᑦ ᖅᐳᒻᒥᓛᓄᑦ, ᐊᑐᐃᓐᓇᖅᑲᐃᓐᓇᕈᔪᑉᐸᓚᐅᖅᒪᑕ ᐃᓚᕇᑦ. ᓱᑯᕌᖕᒐᑕᓕ ᐃᒋᐅᐃᓐᓇᕋᖅᓴᐅᖅᑲᑦᑕᖅᓂᖅᒪᑕ.

Dried urju can be used as a diaper or as a menstrual pad. To make a diaper or pad, Aalasi recalls using cotton flour bags because flour bags were soft and easy to wash. She would tie the flour bag on as a diaper and then stuff the bottom with urju. About two handfuls of urju is sufficient for one diaper. Aalasi recalls that if urju were unavailable, she used rabbit fur instead, but urju was preferable because it was discarded after one use.

ᐋᓚᓯ ᐃᓯᕕᑎᓇᔪᖅ ᐅᖅᔪᖕᓂᒃ ᐃᓕᓚᐅᖏᓐᓂᖅᓂᑦ ᐅᔭᕋᐅᑉ ᖄᖕᒐᓄᑦ ᐸᓂᖅᓯᐊᓇᓗᓂᕆᑦ.

Aalasi separates the *urju* before laying it out on a rock to dry.

Aalasi also remembers using dried urju as bedding for newborn puppies, which her family almost always had. The urju was simply tossed away when it was soiled.

ᐸᓂᖅᑎᓯᒪᔪᑦ ᐅᕐᔪᑦ ᐅᒃᓱᒃᓴᐅᑎᒋᔭᐅᔪᓐᓇᕆᕙᑦ ᑐᐱᕐᓄᑦ ᖃᕐᒪᕐᓄᑦᑐ. ᐊᓇᙱᕐᑲᔾᔭᐃᒃᑯᑎᐅᔪᓐᓇᖅᑐᑎᒡᑐ ᑕᐃᒪ. ᐋᓚᓯ ᐃᖅᑲᐅᒪᒋᕗᖅ ᐊᖑᓇᓱᒃᑎᑦ ᑲᒥᒻᒥᓂᒃ ᐃᑐᑦᑕᖅᓯᐃᖅᑦᑕᕐᓂᕐᒪᑕ ᐅᖅᑰᔾᔪᑎᒃᓴᒃᑲᓐᓂᕐᓂᑦ.

Dried urju can be used around the edges of tents and qarmaqs for insulation. This may also help to keep insects out. Aalasi also recalls hunters adding extra insulation to their kamiks by stuffing dried urju into them.

ᐳᐊᓗᙳᐊᑦ (p. 20) ᐊᒻᒪ ᓱᐳᑎᑦ (ᒥᖅᑯᖓ ᓱᐳᑎᐅᑉ; p. 30) ᐱᑕᖃᙱᑉᐸᑕ ᐅᕐᔪ ᒪᓂᕆᔾᔭᒃᓴᐅᒻᒥᔪᑦ. ᑭᓯᐊᓂ ᐃᒻᒥᒃᑯᑦ ᐃᑭᑦᑐᑎᑦ ᓄᖑᓴᕋᐃᑦᑐᑯᓘᒐᓗᐊᑦ, ᐊᑐᕋᒃᓴᐅᒐᓗᐊᖅᑐᑎᑦ. ᐱᐅᓂᖅᐹᖑᕗᕐᑕ ᐊᑉᐳᐊᖓᓂ ᐃᓚᓗᒍ ᒪᓂᕐᒥᑦ (p. 26).

ᐅᕐᔪ ᐸᓂᖅᑐᕇᓂ ᐅᓐᓄᐊᖏᓐᓇᖅ. ᐅᕐᔪ ᓂᕈᒥᒃᑐᑯᓗᐃᑦ ᐸᓂᖅᑎᓯᒪᓪᓗᒋᑦ ᓲᕐᓗ ᐅᑲᓖᑉ ᒥᖅᑯᖏᑎᑐᑦ.

Urju dried on a rock overnight. When urju is dried, it is as soft as rabbit fur.

If *pualunnguat* (Arctic cotton; p. 20) and *suputit* (*uqpi suputillu*, Arctic willow; p. 30) are not available, urju can be used as a wick for a *qulliq* (soap stone lamp). It is not ideal as a wick because it burns too quickly, but it will suffice. It will burn best if combined, half and half, with *maniq* (lamp moss; p. 26).

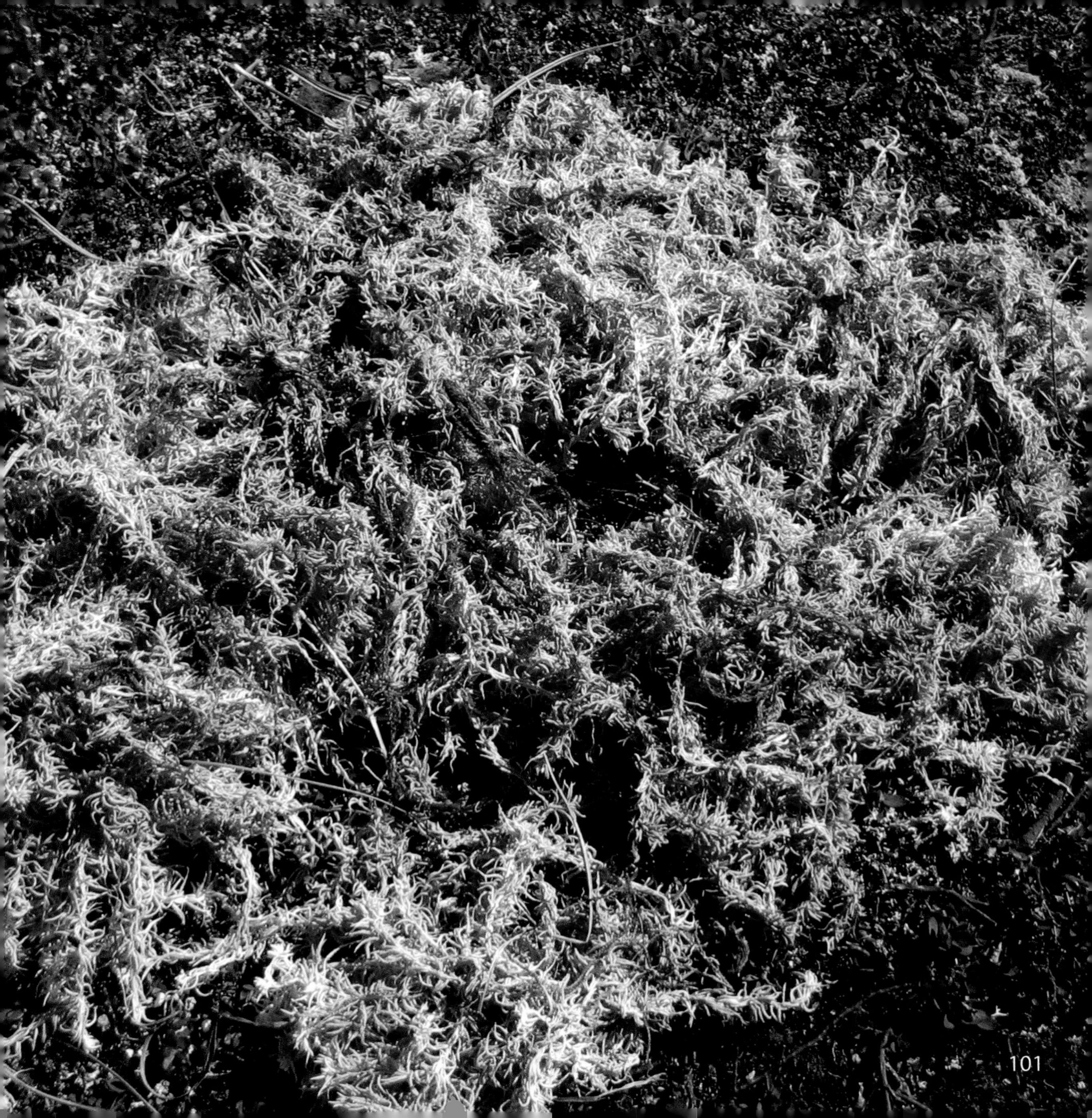

ᐃᔨᓯᐅᑎ
Ijisiuti / River Algae

ᐃᔨᓯᐅᑎ ᐱᕈᖅᑲᑦᑕᖅᑐᑦ ᑰᒪᓛᓂᒃ ᑰᖕᓂᓪᓗ ᓄᓇᕗᒻᒥᑦ. ᐊᒥᐊᖅᑲᖅᐸᑦᑐᑎᑦ ᑐᖑᔪᖅᑕᓂᑦ ᑕᖅᓴᐅᓕᕇᑦᑐᑎᑦ.

Ijisiuti grows on rocks in many creeks and rivers across Nunavut. Its colour ranges from light to dark green and it is made up of many small strands.

ᐃᔨᓯᐅᑎ ᐊᑐᖅᑕᐅᔪᓐᓇᖅᑐᑦ ᐊᑲᐅᓯᕙᐅᑕᐅᓗᑎᒃ ᐃᔨᓄᑦ, ᐃᔨᐅᑉ ᓯᓚᑖᓄᑦ, ᐊᒻᒪ ᖁᕝᕕᐅᕐᕕᓄᑦ. ᐋᓚᓯᐅᑉ ᐊᖓᔪᖓ ᐃᔨᓗᒃᑐᓂ ᒪᒃᑯᑦᑐᓂ. ᐃᔨᖓᓄᑦ ᐃᓕᓯᕙᖅᑲᓚᐅᖅᓯᒪᔪᖅ ᐃᔨᓯᐅᑎᒥᑦ ᐅᓐᓄᐊᓕᒫᖅ ᐊᑕᑎᑕᐅᓪᓗᓂ ᐃᔨᖓᓗ ᖃᓪᓗᓈᖅᑕᒥᑦ ᒪᑐᓪᓗᒍ. ᐃᔨᖓ ᐱᐅᓯᔮᖅᑐᕕᓂᖅ.

Ijisiuti can be used as a medicinal eye ointment to help with eye, eyelid, and tear duct infections. Aalasi's older sister had an eye infection when she was young. Ijisiuti was placed directly on her eyeball, left overnight, and covered with a cloth eye patch. Her eye soon healed.

ᐃᔨᓯᐅᑎ ᐊᑐᖅᑕᐅᔪᓐᓇᕆᕗᑦ, ᐃᔨᓗᒃᑐᓄᑦ. ᓂᑦᑕᓇᑦᑎᐊᖅᑐᐊᓘᕗᑦ ᐊᑲᐅᑦᑐᑎᑦᑐ ᐃᔨᒧᑦ ᐃᓕᔭᐅᑎᑦᑐᒥᑦ. ᐊᑐᖅᑕᐅᔪᓐᓇᕐᒥᔪᑦ ᐃᔾᔨᓯᒪᑐᐃᓐᓇᖅᑐᓄᑦ.

Ijisiuti can also be used to soothe sore, red eyes. It is cool and refreshing when placed on the eyelids. It can also be used to help remove something from the eye.

ᐊᑐᕐᓂᐊᕐᓗᒋᑦ ᐃᔨᓯᐅᑎ, ᐲᔭᓕᕈᕕᒋᑦ ᓴᓂᐅᑦᑕᐃᓕᑎᑦᑎᐊᕋᔭᖅᐸᑎᑦ (ᓯᐅᕋ ᐊᒻᒪ ᒪᕐᕋᕈᕓᖕᒋᓪᓗᑎᕋ). ᐃᔨᓯᐅᑎ ᓴᙱᓂᖅᐹᖑᕐᑲᑦᑕᖅᑐᑦ ᐋᒃᒌᓯᐅᑉ ᓄᙳᐊᓂ ᐊᒻᒪ ᓯᑎᐱᕆᒥᑦ. ᐃᔨᓯᐅᑎᑦ ᑐᕐᑯᓕᖕᒐᖕᒋᑦᑎᐊᕐᓗᑎᒃ ᑭᓯᐊᓂ ᐊᑐᕋᒃᓴᐅᓂᖕᒋᓐᓄᑦ ᐊᐅᔭᒃᑯᑐᐊᖅ ᐊᑐᖅᑕᐅᓃᖕᒍᕙᑦ.

To use ijisiuti, make sure it is clean when you pull it out of the water (no sand or dirt). Choose darker green ijisiuti, as it is riper. Ijisiuti is most potent in late August and September. Ijisiuti should be used fresh from the water and therefore can only be used in the summer.

ᐃᔨᓯᐅᑎᑦ ᐊᑲᐅᓯᓴᐅᑎᐅᕙᓐᓇᖅᑐᑦ ᐃᔨᓗᒃᑐᓄᑦ, ᐊᒻᒪ ᐃᔨᐅᑉ ᑭᓪᓕᖕᒐᓄᑦ ᖁᕝᕕᐅᕝᕕᓄᑦᓗ.

Ijisiuti can help with many ailments of the eye, including irritations of the skin around the eye and infections of the tear duct.

ᓂᕐᓇᐃᑦ
Nirnait / Snow Lichen

ᓂᕐᓇᐃᑦ ᓇᓂᔾᕐᓂᖅᑐᑯᓘᕗᑦ-ᖃᑯᖅᑕᑯᓗᐃᑦ ᓯᓂᖏᑦ ᐃᒧᖅᑳᓯᒪᔪᑯᓗᐃᑦ ᑕᖅᓴᐅᑦᑎᐊᖅᑲᑦᑕᖅᑐᑦ ᑐᖑᔪᖅᑕᓂᒃ ᐱᕈᖅᑐᓂᒃ ᐊᒻᒪᓗ ᐊᓯᖏᖏᓐᓂᑦ ᐱᕈᖅᑐᓂᑦ ᓄᓇᔾᒥᑦ. ᓂᕐᓇᐃᑦ ᐊᑯᓂᐊᓗᒃ ᐱᕈᕋᓱᖅᑲᑦᑕᖅᑐᑦ, ᓱᕐᓗ ᐊᓯᖏᑦ ᑐᒃᑐᓄᑦ ᓂᕿᑦ ᑕᐃᒪᐃᑉᐸᒻᒪᑕ. ᓱᕈᓯᐅᓪᓗᓂᑦ ᐋᓚᓯ ᐅᖃᐅᔾᔭᐅᖃᑦᑕᓚᐅᖅᓯᒪᔪᖅ ᓂᕐᓇᓂᑦ ᐊᓯᖏᓐᓂᓗ ᐱᕈᖅᑐᓂᑦ ᑐᓪᓕᕋᖃᑦᑕᕐᑯᔭᐅᓚᓂ ᐊᑯᓂᐊᓗᒃ ᐱᕈᒃᑲᓐᓂᕋᓱᓲᖑᒻᒪᑕ. ᓂᕐᓇᐃᑦᑕᐅ ᐱᒻᒪᕆᐅᕗᑦ ᑐᒃᑐᓄᑦ ᓂᕿᑭᔭᐅᒻᒪᑕ.

Nirnait is easy to spot—its ivory fingers with crinkled edges stand out from soft green mosses and other plants on the tundra. Like other lichens, nirnait takes a long time to grow. As a child, Aalasi was told to avoid stepping on nirnait, and other plants, to show respect and because it takes so long to regenerate. Also, nirnait is an important food source for caribou.

ᓂᕐᓇᐃᑦ ᐊᑐᖅᑕᐅᔪᓐᓇᖅᑐᑦ ᐊᑲᐅᓯᓴᐅᑎᐅᓗᑎᒃ ᖃᓂᒪᓂᑦ ᐊᓂᐊᑦᑎᔪᓐᓇᕐᒪᑦ ᐊᐅᒫᔾᕐᓂᒃᑯᑦ. ᑏᒻᒎᐊᑦ ᓂᕐᓇᓂᑦ ᓴᓇᓯᒪᔪᖅ ᐊᐅᒫᔾᔪᑕᐅᔪᓐᓇᕐᒪᑦ ᐃᓄᒻᒧᑦ. ᐋᓚᓯ ᐃᖅᑲᐅᒪᔪᖅ ᐊᓈᓇᖓ ᑏᒻᒎᐊᓕᐅᕈᔾᔨᓚᐅᖅᓯᒪᒻᒪᑦ ᖃᓂᒪᔪᒥᒃ ᐅᑭᐅᖑᑎᑦᑐᒍ. ᐋᓚᓯᐅᑉ ᐊᓈᓇᖓ ᐃᓕᒃᑯᐊᓂᑦ ᓂᕐᓇᓂᑦ ᐊᒥᐊᒃᑯᓯᒪᓐᓂᕐᒪᑦ ᐅᑭᐅᖑᑎᑦᑐᒍ. ᑏᒻᒎᐊᖅ ᐃᒥᓚᐅᖅᑐᓂᐅᑉ, ᐊᑯᓂᐅᖏᑦᑐᖅ ᐊᑲᐅᓯᔪᕕᓂᖅ ᑕᐃᓐᓇ ᐃᓄᒃ. ᐊᓯᒃᑲᓐᓂᐊᒍ ᐋᓚᓯ ᐅᑭᐅᖅᑲᖅᑐᓂ 24<ᓗᖕᓂᑦ ᖃᑦᑤᑎᑏᑦ ᖃᓂᒪᓕᓚᐅᖅᓯᒪᓐᓂᕐᒪᑕ, ᐊᑕᖏᖅᑐᑎᒃ ᖃᓂᒪᖏᑦᑐᑑᓕᖅᑐᓂ. ᐃᖅᑲᐅᒪᕗᒥᑦ ᐊᑲᐅᓯᔾᔪᑕᐅᔪᓐᓇᕐᓂᖏᑦ ᓂᕐᓇᐃᑦ, ᑏᒻᒎᐊᓕᐅᕈᑎᓚᐅᖅᓯᒪᔭᕕᓂᖏᑦ ᓂᕐᓇᓂᒃ ᐃᓚᓂ. ᐃᑲᕐᕋᐅᑉ ᐊᕝᕙᑦᑐᐊᖓ ᐊᓂᒍᖅᓯᒪᓕᖅᑎᑦᑐᒍ ᐃᓚᖏᑦ ᐊᐅᒫᔾᖅᑐᐊᓘᓕᖅᑐᕕᓃᑦ.
ᐃᓚᖏᑕ ᐃᓚᖏᑦ ᕿᑲᑐᐃᓐᓇᕈᓐᓃᖅᑐᕕᓃᑦ, ᐃᓚᖏᑦᑐ ᖃᐅᑐᐃᓐᓇᖅᑎᑦᑐᒍ ᐊᑲᐅᓯᒋᑦᑐᑎᑦ.

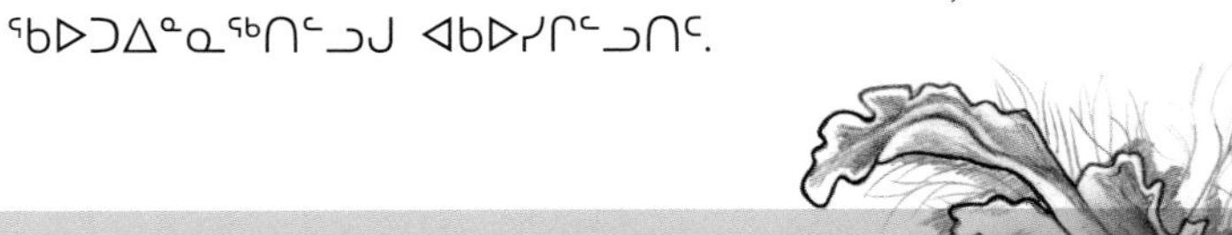

Nirnait can be used as a medicine for illnesses that need to be sweated out. Nirnait tea causes a person to sweat. Aalasi recalls that her mother once made nirnait tea for someone who was sick in the winter. Aalasi's mother had saved a small parcel of nirnait for the winter. Soon after drinking the tea, the person became better. Another time, when Aalasi was about 24 years old, her entire family became very sick except for her. She remembered the healing properties of nirnait, so she served her family nirnait tea. About half an hour after drinking the nirnait tea, they began to sweat profusely. Some of them became active the following day and the others were feeling better soon after.

ᓂᕐᓇᐃᑦ ᓂᕆᔭᐅᔭᕆᐊᖃᙱᓪᓚᑦ. ᑏᙵᐊᓕᐊᕆᓗᒍ ᑭᓯᐊᓂ ᐱᐅᒻᒪᑕ. ᓂᕐᓇᓂᒃ ᑏᙵᐊᓕᐅᕐᓂᐊᕈᕕᑦ, ᑎᖅᑎᑎᑲᐃᓐᓇᕆᐃᓐᓇᕐᓗᒍ ᓈᒻᒪᒃᑲᔭᖅᑐᖅ, ᐃᓱᖅᓯᓇᓱᕆᐃᓐᓇᕐᓂᖓᓂᑦ. ᐃᒥᕐᓂᐊᕆᐃᓐᓇᕐᓗᒍ ᑎᓴᐃᖓᓕᖅᐸᑦ. ᓂᕐᓇᐃᑦ ᖃᓵᔾᓇᑯᓗᐃᑦ ᑏᓕᐅᕈᑎᒥᑦ ᑕᑕᐃᔦᓐᓇᖅᐳᑦ. ᐊᒻᒪᓗ ᐃᕝᙳᓯᖅ ᐊᑕᐅᓯᖅ ᓈᒻᒪᕆᐃᓐᓇᖅᑐᓂ. ᓂᕐᓇᐃᑦ ᑏᙵᐊᖑᓪᓗᒍ ᓴᙱᔦᕿᔫᒻᒪᑦ, ᓄᑕᕋᓄ ᐱᔭᐅᔭᕆᐊᖃᖕᒋᑦᑐᖅ. ᓂᕐᓇᐃᑦ ᖃᖓᕆᐃᓐᓇᑦᑎᐊᖅ ᓄᐊᑕᒃᓴᐅᙳᑦ, ᑭᓯᐊᓂᓕ ᐆᖅᓯᒪᓗᒋᑦ ᖃᐅᒪᔦᑭᔮᒥᑦ ᐱᐅᑦᑎᐊᖅᑐᑦ. ᓂᕐᓇᐃᑦ ᓂᑲᒋᔭᐅᔭᕆᐊᖃᖅᑐᑦ, ᖃᓪᓗᓈᖅᑕᒦᓪᓗᑎᒃ, ᐊᑭᓯᐅᓪᓘᓐᓃᑦ ᐆᖓᓂᑦ ᓂᓪᓚᓱᑦᑐᒦᓪᓗᑏᑦ.

ᓂᕐᓇᐃᑦ ᐃᒎᑦᑰᔭᖅᓯᒪᔦᑯᓗᐃᑦ ᑕᖅᓴᐅᑦᑎᐊᖅᑐᑦ, ᓰᕐᓗᓕ ᐊᐳᑎ ᓄᓇᔭᒥᑦ.

The bright, crinkled fingers of *nirnait* resemble a patch of snow on the tundra.

Nirnait should not be eaten. It should only be consumed as a tea. To make nirnait tea, simply boil the nirnait briefly, until the water becomes dark. Drink it when it has cooled. Only a small handful of nirnait is needed to make a pot of tea and one cup of the tea is enough to be effective. Because nirnait tea is so powerful, it probably should not be given to babies. Nirnait can be collected at any time, but it is best not to store it in plastic. It should be handled gently and stored in a flour sack, some canvas, or in something like a pillowcase.

ᓂᕐᓇᐃᑦ ᐊᑯᓂᐊᓗᒃ ᐱᕈᕋᓱᓲᑦ. ᐋᓚᓯᓕ ᐃᖅᑲᐅᒪᔫᖅ ᑐᑦᓕᕋᖅᑲᑦᑕᖅᑯᔭᐅᑎᓐᓇᒥᑦ ᐱᔮᓇᓗᓂᑦ.

Nirnait takes a long time to grow, so Aalasi was always told to avoid stepping on it, she recalls.

ᓂᕐᓇᐃᑦ ᖃᑯᑦᑎᐊᖅᑐᐊᓘᖅᑲᑦᑕᖅᑐᑦ ᓯᖅᑭᓂᖅ ᓂᐱᓕᖅᑎᑦᓗᒍ.

Nirnait appears ivory-coloured in the low light of dusk.

ᐊᐅᔭᐅᑉ ᓄᖑᐊᓂ ᐃᓯᕆᐅᑕᐅᔭᖅ (ᐳᐱᒃ) ᓂᒃᑯᖕᓂᐊᖅᑐᖅ.
ᓄᐊᑕᐅᔪᓐᓇᖅᑐᑦ ᐃᓅᓕᓴᐅᑎᒋᔭᐅᓂᐊᕐᓗᑎᒃ.

This live mushroom will dry out as summer ends.
Then, it can be harvested and used for medicinal purposes.

ᐳᔪᐊᓗᒃ
Pujualuk / Dried Mushroom

ᐳᔪᐊᓗᒃ (ᐸᓂᖅᓯᒪᔪᖅ ᒪᔅᕉᒻ) ᓇᓂᔭᕐᓂᖕᒋᑦᑐᑦ ᓲᖅᑲᐃᒻᒪ ᑕᖅᓴᐅᖕᒋᑦᑐᑯᓗᐃᑦ ᖁᖅᓯᐸᓗᒃᑐᑎᒃ ᑲᔪᐃᑦ ᑕᖅᖃᓕᓂᖅᒦᒡᒍᔪᐊᓘᑦᓗᑎᑦᓗ ᐱᕈᖅᑐᐃᑦ ᑕᑭᔪᐃᑦ ᐅᕝᕙᓘᓐᓃᑦ ᐅᔭᖃᐃᑦ ᓴᓂᖏᓐᓂ. ᐳᔪᐊᓗᒃᓯᐅᖅᓂᐊᕈᕕᑦ, ᕿᓂᓇᐊᖅᑲᖅᑐᑎᑦ ᓯᐅᖃᔮᒥᐅᖕᒋᑐᖅ, ᓄᓇᔭᒥ ᑭᓯᐊᓂ.

It is not easy to spot a *pujualuk* because they are generally dull beige in colour and they tend to appear in the shadows of taller plants and rocks. To find a pujualuk, look in areas without gravel where the soil is richer.

ᐳᔪᐊᓗᒃ ᐊᑑᑎᓚᒻᒪᓈᑦ "ᒪᑦᑐᑎᑦᑎᐊᕚᓘᑦᓗᑎᑦᓗ". ᓱᑭᔾᔭᐃᒃᑯᑎᑦᑎᐊᕚᓗᐃᑦ ᐊᒻᒪᓗ ᐊᐅᓈᖅᑐᒥᑦ ᓯᒥᑦᑐᐃᑦᑎᐊᖅᑐᑎᒃ. ᐳᔪᐊᓗᒃ ᐊᒥᖅᓗᑭᔪᓐᓂᒃ ᐱᐅᓯᕚᑦᓕᖅᑎᑦᑎᔪᓇᑦᑎᐊᖅᒥᔪᑦ, ᓲᖅᓗ ᐅᕕᓂᖅᓗᖕᓂᑦ ᐊᒻᒪᓗ ᐊᓯᑭᔫᔭᖕᒋᓐᓂᑦ. ᐳᔪᐊᓗᓐᓂᒃ ᐊᑐᖅᓂᐊᖅᓗᓂ, ᑭᑦᓚᒧᑦ ᐅᕝᕙᓘᓐᓃᑦ ᐊᒥᖅᓗᒧᑦ ᓱᖃᑐᐃᓐᓇᖅᓗᒍ ᐃᓕᔪᓐᓇᖅᑕᐃᑦ ᐳᔪᑭᔫᓂᖕᒐ ᐅᕙᓂᖕᓄᑦ ᐊᑦᑐᐊᖕᒎᖅᓗᓂ. ᖃᐅᔨᒪᔭᐅᔭᕆᐊᓕᒃ: ᐳᔪᐊᓗᒃ ᐸᓂᖅᓯᒪᔪᖅ ᑭᓯᐊᓂ ᐱᐅᕗᖅ ᐸᓂᖅᓯᒪᖕᒋᒃᑯᓂ ᐊᑲᐅᓯᓴᐅᑎᖃᖕᒋᒻᒪᑦ.

A pujualuk can be very useful as a "band-aid." It will protect the wound and it contains vital blood-clotting chemicals that can help to stop bleeding. A pujualuk can also alleviate general skin problems, such as rashes and other irritations. To use a pujualuk, simply break it open and place it powdery-side down on the wound or skin irritation.

ᐳᔪᐊᓗᒃ ᐊᐅᔭᓕᒫᖅ ᓄᐊᑕᒃᓴᑦ ᐊᒻᒪ ᐃᓕᒃᑯᐊᓯᔭᒃᓴᐅᔪᑦ ᐱᐅᔪᓐᓃᕈᓐᓇᕋᑎᒃ.

Note: live mushrooms do not have the medicinal properties that dried mushrooms have.

ᐋᓚᓯ ᐃᖅᑲᐅᒪᔪᖅ ᐊᓈᓇᖓ ᐆᖅᓯᒪᔪᓂᒃ ᐳᔪᐊᓗᖅᑲᐃᓐᓇᖅᑲᑦᑕᓚᐅᖅᓯᒪᒻᒪᑦ ᑭᓚᖅᑐᖅᑲᖅᓂᕈᓂ ᓇᑦᓚᐅᒃᑯᒫᓇᑦᑐᓂᒋᑦ. ᐊᖓᖖᒐᐃᑦ ᐊᒥᖓᓂᖕᖏᓐᓂᑦ ᐆᖅᑲᖅᑎᒡᒍᖅᑲᑦᑕᓚᐅᖅᓯᒪᔭᖕᖏᑦ ᐱᔾᔪᑎᒋᑦᑐᒍ ᐊᖓᖖᒐᐃᑦ ᐊᒥᖕᖏᑦ ᓂᕈᒥᒃᑐᑯᓘᒻᒪᑕ ᓱᕋᑦᑎᓇᖕᖏᒻᒪᑕ ᓂᑲᓇᖅᑐᓂᑦ ᐳᔪᐊᓗᓐᓂᑦ. ᐳᔪᐊᓗᓐᓂᑦ ᐱᓂᐊᕈᖓᑦ ᖄᖕᒐ ᓱᕋᑦᑕᐃᓚᒪᑦᑎᐊᖅᑐᒍ ᐳᔪᐊᓘᓂᖓ ᓴᒐᑐᐃᓐᓇᑦᑕᖅᓂᐊᖅᒪᑦ.

A pujualuk can be collected during the summer months and stored indefinitely. Aalasi recalls that her mother always kept a pouch of them handy just in case someone was injured. She preferred to store them in a lemming pelt bag because the padding and softness of the lemming pelt would protect the fragile pujualuk. When picking a pujualuk, try not to rupture the outer skin any more than necessary so that the powdery insides stay intact.

ᐋᓚᓯ ᐳᔪᐊᓗᒻᒥᑦ
ᒪᑉᐱᕐᓯᔪᕐ
ᖃᓄᖅ ᑭᓪᓚᓂᑦ
ᐊᑐᖅᑕᐅᓲᖑᒻᒪᖔᑕ
ᑕᑯᑎᑦᑎᓂᐊᕐᒥᒃ.

Aalasi opens a
powdery *pujualuk*
and demonstrates how
she would place it on a
wound.

ᐊᑐᒐᒃᓴᒃᑲᓐᓂᑦ
Additional Resources

ᑐᑭᓯᒋᐊᒃᑲᓐᓂᕈᒪᒍᕕᑦ ᓄᓇᕋᐃᑦ ᐅᖃᓕᒫᒐᕐᒥᑦᑐᑦ ᒥᒃᓵᓄᑦ, ᐅᑯᐊ ᐊᑐᕈᒪᑐᐃᓐᓇᕆᐊᓕᒃᑎᑦ.

If you would like more information about the plants in this book, you may want to use the following resources.

- ᓄᓇᕗᒻᒥᑦ ᐱᕈᖅᑐᐃᑦ. 2004. ᑭᐅᕈᓕᓐ ᒫᓗᕆ, ᓲᓴᓐ ᐊᐃᑲᓐ. ᓄᓇᕗᑦ ᐃᓕᓐᓂᐊᖅᑐᓕᕆᔨᒃᑯᑦ, ᐃᖃᓗᐃᑦ, ᓄᓇᕗᑦ. (ᖃᑉᓗᓈᑎᑐᑦ ᐃᓄᒃᑎᑐᑦᓗ)

  Common Plants of Nunavut. 2004. Carolyn Mallory and Susan Aiken. Nunavut Department of Education, Iqaluit, Nunavut. (English and Inuktitut)

- ᓄᓇᔾᒥᑦ ᐱᐅᔦᐃᑦ: ᐱᕕᐊᕐᓇᖅᑐᑦ ᓄᓇᕋᐃᑦ ᑲᓇᑕᐅᑉ ᐅᑭᐅᖅᑕᖅᑐᖕᒐᓂᑦ. 2004 (ᓴᓇᔾᐅᒋᐊᒃᑲᓐᓂᖅᓯᒪᔪᑦ). ᐸᐃᔅ ᗷᑦ. ᐊᑉ Hᐃᐅ ᓴᖅᑭᑎᑦᑎᔾᔪᑦ, ᔾᓗᓇᐃᕝ, ᓄᓇᑦᓯᐊᖅ. (ᖃᑉᓗᓈᑎᑐᑦ ᐊᒻᒪᓗ ᓇᐃᓪᓕᑎᖅᓯᒪᔦᓪᓚᑦ ᐃᓄᐃᓐᓇᖅᑐᑦ)

  Barrenland Beauties: Showy Plants of the Canadian Arctic. 2004 (revised edition). Page Burt. Uphere Publishing, Yellowknife, Northwest Territories. (English with summaries in Inuinnaqtun)

- ᖃᐅᔨᓇᓲᑎᑦ ᐊᑯᓐᓂᑦᑎᓐᓂᑦ ᐅᑭᐅᖅᑕᖅᑐᒥ. 1994. ᐃ.ᓯ. ᐸᐃᓗ. ᓯᑕᑐᖅᓴᖅᕕᔾᔦᐊᒥᑦ ᓰᑲᒍ ᐳᕋᔅ, ᓰᑲᒍ, ᐃᓕᓄᐃ. (ᖃᑉᓘᓈᑑᖅᑐᑦ)

  A Naturalist's Guide to the Arctic. 1994. E.C. Pielou. The University of Chicago Press, Chicago, Illinois. (English only)

- ᐊᐱᖅᓱᕐᓂᖅ ᐃᓄᐃᑦ ᐃᓐᓇᖅᑭᓐᓂᑦ: ᖃᓄᐃᑦᑕᐃᓕᒪᓇᓲᑎᑐᖃᐃᑦ ᑎᒥᒧᑦ. 2001. ᐃᓕᓴᐱ ᐆᑦᑐᕙ, ᑎᐴᓚ ᖄᐱᒃ ᐊᑕᒍᑦᓯᐊᖅ, ᑎᕆᓯ ᐃᔾᔭᖏᐊᖅ, ᔭᐃᑯ ᐱᑦᓯᐅᓚᒃ, ᐋᓚᓯ ᔫᒥ, ᐊᑭᓱ ᔫᒥ, ᒪᓚᐃᔭ ᐸᐹᑦᓯ. ᐊᖅᑭᒋᐊᖅᓯᔨ ᒥᓯᐅᓪ ᑐᕆᐃᓐ ᐊᒻᒪ ᕗᕆᑦᑐᕆᒃ ᓚᒍᕋᓐ. ᓄᓇᕗᑦ ᓯᓚᑦᑐᖅᓴᕐᕕᒃ, ᐃᖃᓗᐃᑦ, ᓄᓇᕗᑦ. (ᖃᑉᓗᓈᑎᑐᑦ ᐃᓄᒃᑎᑐᓗ)

  Interviewing Inuit Elders: Perspectives on Traditional Health. 2001. Ilisapi Ootoova, Tipuula Qaapik Atagutsiak, Tirisi Ijjangiaq, Jaikku Pitseolak, Aalasi Joamie, Akisu Joamie, and Malaija Papatsie. Edited by Michèle Therrien and Frédéric Laugrand. Nunavut Arctic College, Iqaluit, Nunavut. (English and Inuktitut)

- ᐊᑲᐅᓯᓴᐅᑎᑐᖅᑲᓕᕆᓂᖅ: ᐅᓂᒃᑳᓕᐊᖑᓯᒪᔪᑦ, ᓰᑎᐱᕆ *28, 1983*.1984. ᔫᓇᑕᓂ ᓯᑏᕝ. ᐊᕙᑕ ᐃᓕᖅᑯᓯᓕᕆᓂᖅ, ᐃᓄᒃᔪᐊᖅ, ᑯᐸᐃᒃ. (ᖃᑉᓗᓈᑎᑐᑦ ᐃᓄᒃᑎᑐᓗ)

  Traditional Medicine Project: Interim Report, September 28, 1983. 1984. Jonathan Stevens. Avataq Cultural Institute, Inukjuaq, Quebec. (English and Inuktitut)

ᑎᑎᕋᖅᑐᕕᓂᑦ

Contributors

ᐋᓚᓯ ᔫᒥ ᐃᓅᓂᑯ ᐃᓄᒃᔪᐊ, ᑯᐸᐃᒻᒥᑦ. ᖃᑕᙳᑎᖏᑦ ᓄᑦᑎᕐᓂᑯᐃᑦ ᐸᓐᓂᖅᑑᒧᑦ ᓂᕕᐊᖅᓯᐊᙳᑎᓪᓗᒍ. 1960ᒥᑦ, ᓄᑦᑎᕐᓂᑯ ᐅᐃᒌᑦ ᕿᑐᕐᖓᓈᑦ ᓂᐊᖁᙴᒧᑦ ᐃᓪᓗᑖᕆᐅᖅᑐᑎᓪᓗ. ᑕᐃᒪᙵᓂ ᓂᐊᖁᙴᒥᐅᑕᐅᔪᖅ. ᐊᕐᕌᒍᓗᒐᓚᖕᓂᑦ, ᓇᔾᔨᔦᓕᕆᔨᐅᖅᑲᑕᐅᓚᐅᖅᓯᒪᔪᖅ ᐋᓐᓂᐊᕕᒻᒥᑦ. ᐋᓚᓯ ᐊᐱᖅᓱᖅᑕᐅᖅᑲᑕᐅᓂᑰᒻᒥᔪᖅ ᐃᓐᓇᖅᓂᑦ ᐅᖅᑲᓕᒫᒐᖅᒧᑦ ᖃᓄᐃᑦᑕᐃᓕᒪᔾᔪᑎᑐᖅᑲᐃᑦ ᑎᒥᒧᑦ ᐊᒻᒪ ᐃᓕᓐᓂᐊᖅᑎᑎᓲᖑᑦᑐᓂ ᓄᓇᕋᐃᑦ ᒥᒃᓵᓄᑦ ᓯᓚᑦᑐᖅᓴᕐᕕᒻᒥᑦ. ᐊᐅᓪᓚᖅᑲᑕᐅᕙᑦᑐᓂᓗ ᑲᓇᑕᒥ ᓯᓚᑖᓄᓗᑦ ᓄᓇᕋᐃᑦ ᐊᑑᑎᖏᑦ ᐅᓂᒃᑳᓇᔭᖅᑐᖅᑐᕆᑦ.

Aalasi Joamie was born in Inukjuak, Quebec. Her family moved to Pangnirtung when she was a young girl. In the 1960s, she moved to Niaqunnguuq (Apex) with her husband and children into their first house. She has lived there ever since. For many years, Aalasi worked as a maternity aid at Baffin Regional Hospital. Aalasi contributed to *Interviewing Inuit Elders: Perspectives on Traditional Health* and she teaches traditional plant knowledge workshops at Nunavut Arctic College. She also travels to traditional plant-use conferences nationally and internationally.

ᐋᓇ ᓯᒡᓗᕐ ᐃᖅᑲᓗᒥᐅᑕᐅᔪᖅ, ᓯᓚᑦᑐᖅᓴᕐᕕᒻᒥᑦ ᐃᓕᓴᐃᔨᐅᖅᑲᑦᑕᖅᑐᖅ ᐊᒻᒪᓗ ᐃᓕᓐᓂᐊᒐᒃᓴᓂᑦ ᑲᒪᔨᐅᓪᓗᓂ. ᑎᑎᕋᖅᑕᐅᓯᒪᔪᖅ ᑐᑭᓯᒋᐊᕈᑎ ᖃᐅᔨᓴᖅᑐᓕᕆᓂᕐᒧᑦ ᐋᓚᓯᒥᑦ ᑕᑯᔭᕆᐅᓚᐅᖅᓯᒪᔪᖅ ᒥᖅᓱᖅᑎᑦᑎᓂᕆᖅᑲᑦᑕᖅᑕᖏᓐᓄᑦ ᐊᑭᖅᑲᖏᑦᑐᓄᑦ ᐅᐸᖅᑲᑕᐅᒐᒥ.

Anna Ziegler lives in Iqaluit, where she works at Nunavut Arctic College as an instructor and regional program coordinator. She is the co-author of *Tukisigiaruti Qaujisaqtulirinirmut: A Life Sciences Handbook for Nunavut Educators.* She first met Aalasi at Aalasi's famous free community sewing lessons.

ᓈᐱᑲ ᕼᐊᐃᓐᓄ, ᐃᓕᓴᐃᔨᒃᓴᐅᓂᖅᒧᑦ ᐃᓕᓐᓂᐊᓈᕋᑖᐸᓗᒃᓯᒪᔪᖅ, ᑲᖏᖅᑑᒐᐱᒻᒥᑦ ᐃᓕᓴᐃᔨᐅᔪᖅ ᐊᑦᑎᓐᓂᖅᓴᓄᑦ. ᑎᑎᕋᖅᑕᐅᓯᒪᔪᖅ ᐃᓕᓐᓂᐊᖅᑐᓕᕆᓂᕐᒧᑦ ᑐᕌᖓᔪᓂᑦ ᐃᓕᓴᐃᔨᒃᓴᓄᑦ ᐃᓕᓐᓂᐊᖅᑐᒃᑯᑦ ᐊᒻᒪ ᓄᓇᕗᒻᒥᑦ ᒪᖀᓐᓂᑦ ᐅᖅᑲᐅᓯᓕᕆᔨᒃᑯᑦ, ᐱᖅᑲᓯᐅᑎᓪᓗᒋᑦ ᓴᖅᑭᑕᐅᓂᑯᑦ

ᓈᓴᐅᑎᓕᕆᔾᔪᑎᑦ ᓄᓇᕗᑦ ᐃᓕᓐᓂᐊᕐᕖᖏᓐᓄᑦ: ᖁᑦᑎᓐᓂᓕᓪ 1, 2 ᐊᒻᒪ 3, ᐊᒻᒪᓗ ᓇᓗᓇᐃᖅᐃᓂᖅ ᖃᒥᖅᓗᓕᓐᓂᑦ.

Rebecca Hainnu, a recent graduate of the Nunavut Teacher Education Program, lives in Clyde River where she works as an elementary school teacher. She has worked on several educational publications with the Nunavut Teacher Education Program and the Nunavut Bilingual Education Society, including *Math Activities for Nunavut Classrooms: Grades 1, 2, and 3* and *Classifying Vertebrates.*